The Sassafras Guide to Zoology

WRITTEN BY PAIGE HUDSON

THE SASSAFRAS GUIDE TO ZOOLOGY

Third Edition, First Printing
Copyright @ Elemental Science, Inc.
Email: support@elementalscience.com

ISBN # 978-1-935614-22-7

Printed In USA For worldwide distribution

For more copies write to :
Elemental Science
PO Box 79
Niceville, FL 32588
support@elementalscience.com

COPYRIGHT POLICY

QUICK START GUIDE

Welcome to your super, scientific journey with the Sassafras Twins!! The information and activities in this guide will help you turn a simple adventure novel into a complete science program for your elementary students. Let's start by answering three pressing questions!

WHAT WILL WE LEARN?

Students will learn about zoology, which is the study of animals. See p. 11 for a list of the topics explored in this program.

WHAT DO I NEED?

In addition to this activity guide, you will need the following materials:

1. **Novel** – All the main reading assignments are from *The Sassafras Science Adventures Volume 1: Zoology*. You can get the paperback novel, the Kindle version, or the audiobook.

2. **Student Materials** – You can have your students use a blank notebook or you can purchase *The Official Sassafras SCIDAT Logbook: Zoology Edition* for each student. Get a glimpse of this option on p. 7. (SCIDAT stands for scientific data and it comes from the Sassafras Twins' journey.)

3. **Demonstration Supplies** – See p. 12 for a full list, or save yourself time and get the *Sassafras Science Year 1 Experiment Kit*, which includes the materials for both volume 1 and volume 2.

If you want more information than what is already in the novel, the following encyclopedias are scheduled in this guide:

- *First Encyclopedia of Animals, Kingfisher First Reference* (best for grades K through 3)
- *DK Encyclopedia of Animals* (best for grades 3 through 6)

If you want to add more fun with optional STEAM* projects, you can find a list of the project supplies on p. 13.

*STEAM: Science, Technology, Engineering, Art, and Math

WHAT WILL A WEEK LOOK LIKE?

Each week you and your students will:

- **Read** scientific information from an adventure-filled novel, also known as a living book, and discuss what you read.

- **Write** down what the students have learned and seen in a way that is appropriate for their skills by keeping a notebook, or rather a SCIDAT Logbook.

- **Do** hands-on science through demonstrations using the directions found in this guide.

You can also add in the optional copywork, library books, and STEAM projects if you want to dig deeper into a topic. For a more detailed explanation of the components in each lesson, we highly recommend checking out the peek inside this guide on pp. 6-7 and reading the introduction on pp. 8-10. The chapter lessons begin on p. 17.

THE
SASSAFRAS SCIENCE
ADVENTURES

The Sassafras Guide to Zoology
Table of Contents

Activity Guide At-A-Glance

> *Take an adventure-filled journey to learn about science!*

1. & 2. Scheduling Options

Choose from a grid-style schedule (1) or a list-style schedule (2). Either way, these scheduling options will make planning your weekly science adventure a snap! These schedule sheets include a summary of the chapter in case your students are reading the novel or listening to the audiobook on their own.

READ

3. Reading Assignments

Know what to read each week in the corresponding Sassafras Science novel. Plus, get options for additional encyclopedia pages to read and for books to check out from the library. The novel contains the essential information for each week, but if you want to dig deeper, we've got you covered!

WRITE

4. SCIDAT Logbook Info

Have confidence that your students are grasping the key points from the reading with the information in the notebooking section. Here, you will find the scientific details that were shared in the chapter, which could be included in your students' narrations or list of facts.

5. Relevant Vocabulary

Build your students' science vocabulary with words relevant to the topics the students are studying.

6. Copywork

Use these selections as memory work, copywork, or dictation—it's up to you!

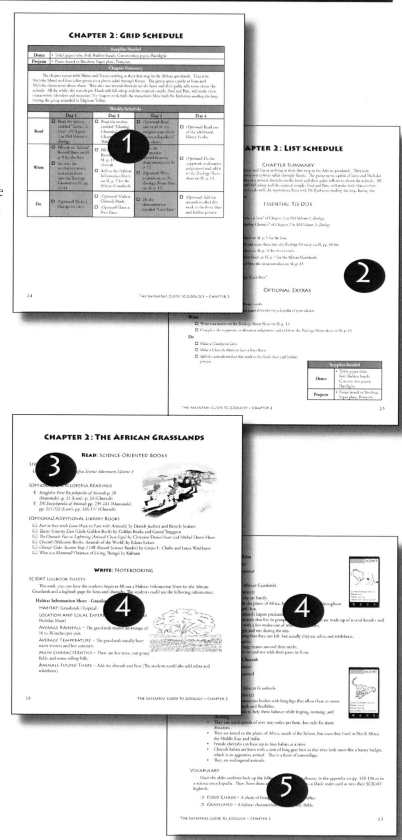

ACTIVITY GUIDE AT-A-GLANCE

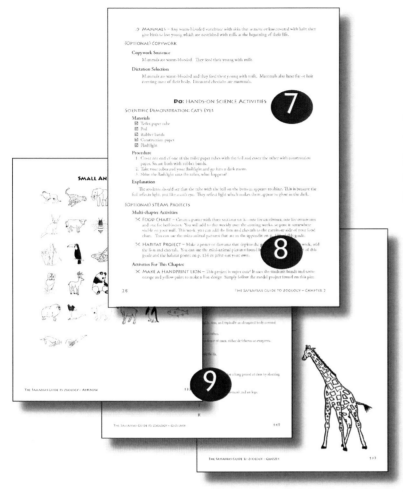

DO

7. RELATED SCIENTIFIC DEMONSTRATIONS

Know what materials you will need to do a weekly hands-on science activity that coordinates with the topic. This section lists the supplies you will need, provides easy-to-follow steps, and explanations to make it a snap to complete the scientific demonstration.

8. COORDINATING STEAM* ACTIVITIES

Add in a bit of STEAM with these optional activity ideas. You will find ideas for projects that last throughout the novel and ones specific to the chapter (week) you are on.

9. TEMPLATES AND MORE

In the guide's appendix, you will find templates for the projects, a full glossary, and a set of quizzes to use along the journey.

*STEAM: Science, Technology, Engineering, Art, and Math

THE SCIDAT* LOGBOOK

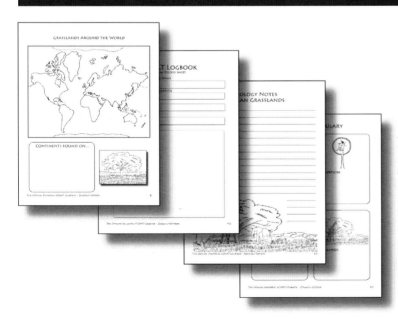

> *Don't forget the SCIDAT logbook for your students!!*

The SCIDAT logbook will serve as a record of your students' journey! It contains all the pages the students will need as they follow like Blaine and Tracey. Each page has been attractively illustrated for you so you don't have to track down pictures for the students to use! Get it all at:

https://elementalscience.com/collections/ sassafras-science

*SCIDAT: Scientific Data

THE SASSAFRAS GUIDE TO ZOOLOGY
INTRODUCTION

Our living books method of science instruction was first proposed in *Success in Science: A Manual for Excellence in Science Education*. This approach is centered on living books that are augmented by notebooking and scientific demonstrations. The students read (or are read to) from a science-oriented living book, such as *The Sassafras Science Adventures Volume 1: Zoology*. Then, they write about what they have learned and complete a related scientific demonstration or hands-on project. If time and interest allow, the teacher can add in non-fiction books that coordinate with the topic, do an additional activity, or memorize related information. If you want to learn more about how this works, you can listen to this free conference session on using living books for science:

☞ Inspiring your students to love science through living books: https://youtu.be/Dvk1LfYGONw

The books in *The Sassafras Science Adventures* series are designed to give you the tools you need to employ the living books method of science instruction with your elementary students. For this reason, we have written an activity guide and a logbook that corresponds to each novel. This particular activity guide contains 18 chapters of activities, reading assignments, scientific demonstrations, and so much more for studying zoology.

Each of the chapters in this guide corresponds directly to the chapters in *The Sassafras Science Adventures Volume 1: Zoology*. They were written to give you the information you need to turn the adventure novel into a full science course for your elementary students. They will provide you with a buffet of options you can use to teach the students about animals. So pick and choose what you know you and your students will enjoy!

WHAT EACH CHAPTER CONTAINS

Each chapter begins with your two scheduling options—a grid schedule and a list schedule. These contain a summary of the corresponding chapter in *The Sassafras Science Adventures Volume 1: Zoology* and the same weekly assignments, but in a different format. These schedules are included to give you an idea of how your week could be organized, so please feel free to alter them to suit your needs. Following the scheduling options, you will find the details for reading, writing, and doing science for the particular chapter. This information is divided into the following sections:

READ: GATHERING INFORMATION

ⓘ LIVING BOOK READING ASSIGNMENT – This section contains the corresponding chapter in *The Sassafras Science Adventures Volume 1: Zoology*.

⚲ (OPTIONAL) ENCYCLOPEDIA READINGS – This section contains possible reading assignments from:
- *First Encyclopedia of Animals, Kingfisher First Reference* (best for grades K through 3)
- *DK Encyclopedia of Animals* (best for grades 3 through 6)

You can choose to read them to the students or have the students read them on their own.

📖 (OPTIONAL) ADDITIONAL LIBRARY BOOKS – This section contains a list of books that coordinate with what is being studied in the chapter. You can check these books out of your local library.

WRITE: KEEPING A NOTEBOOK

☐ SCIDAT LOGBOOK INFORMATION – This section has the information that the students could have included in their SCIDAT logbooks. (SCIDAT stands for scientific data and it comes from the Sassafras Twins' journey.) The students may or may not have the same information on their logbook sheets, which is fine. You want their SCIDAT logbooks to be a record of what they have learned. The logbook information is included as a guide for you to use as you check their work. For more information about notebooking, please read the following article:

  ✎ What is notebooking? – https://elementalscience.com/blogs/news/what-is-notebooking

  ✎ How to use notebooking with different ages – https://elementalscience.com/blogs/news/note-booking-with-different-ages

✐ VOCABULARY – This section includes vocabulary words that coordinate with each chapter. If the students are older, we recommend that you have them create a glossary of terms using a blank sheet of lined paper or the glossary sheets provided in *The Official Sassafras Student SCIDAT Logbook: Zoology Edition*. You can also have them memorize these words and their definitions.

☞ (OPTIONAL) COPYWORK – This section contains a short copywork passage and a longer dictation passage for you to use. Some students may use the shorter passages for dictation or the longer passages for copywork. Feel free to tailor the selections to your students' abilities. You can also use the selections as memory work assignments for the students.

⧗ (OPTIONAL) QUIZ – This section contains the answers for the quizzes included in the appendix. These simple, short quizzes are optional. You can use them as graded quizzes or as review sheets.

DO: PLAYING WITH SCIENCE

☑ SCIENTIFIC DEMONSTRATION – This section includes a list of materials, the instructions, and an explanation for a scientific demonstration that coordinates with the chapter. There is a blank lab report sheet provided for you in the appendix on pp. 127-128 if you would like the students to do a write-up of the demonstration. If the students are in grade 4 or higher, we recommend that they complete at least one of these activities for this course.

✂ (OPTIONAL) STEAM* PROJECTS – These sections contain additional STEAM projects and activities that correspond to the topics in the chapter. There are multi-chapter activities that students can do over the course of several chapters or over the full novel. Plus, there are activities that coordinate with each specific chapter. Pick and choose the activities that interest you and your students.

*STEAM: Science, Technology, Engineering, Art, and Math

ADDITIONAL MATERIALS

We have provided a few additional materials in the back of this guide for your convenience. First, you will find the templates you need for the projects suggested in this guide. Next, you will find a glossary of terms, which you can use with the students as they define the words for each chapter. And finally, you will find a set of eight simple quizzes you can use with the students to verify they are retaining the material.

Quick Links

View all the links mentioned in this guide in one place and get a digital copy of the templates, glossary, and quizzes by visiting the following page:

🖱 https://elementalscience.com/blogs/resources/volume-1-links

For the Students

The SCIDAT logbook is meant to be a record of the students' journey through their study of zoology. It is explained in more detail in Chapter 1 of this guide. You can choose to make your own or purchase a premade logbook from Elemental Science. *The Official Sassafras SCIDAT Logbook: Zoology Edition* has all the pages the students will need to create their own logbook. Each page has been attractively illustrated for you so you don't have to track down pictures for the students to use. This way they can focus on the information they are learning.

Final Thoughts

As the author and publisher of this curriculum I encourage you to contact me with any questions or problems that you might have concerning *The Sassafras Guide to Zoology* at support@elementalscience.com. I, or a member of our team, will be more than happy to assist you. I hope that you and your students enjoy your journey through zoology with the Sassafras twins!

~ Paige Hudson

TOPICAL LIST

The Sassafras Science Adventures Volume 1: Zoology covers a variety of aspects of zoology, such as:

- Habitats
- Classification
- Animal Diet
- Life Cycles (Frog and Butterfly)
- Vertebrates and Invertebrates
- Migration
- Basic Mapping Skills
- Animal Defenses

In the process, you will learn about the following specific animals:

- Lion
- Cheetah
- Elephant
- Giraffe
- Camel
- Cobra
- Spiny-tailed Lizard
- Fennec Fox
- Cow
- Bee
- Chicken
- Spider
- Sloth
- Toucan
- Poison Dart Frog
- Blue Morpho Butterfly
- Koala
- Rabbit
- Panda
- Golden Eagle
- Powerful Owl
- Sambar Deer
- Golden-haired Monkey
- Mice
- Musk Ox
- Snow Goose

- Polar Bear
- Mountain Goat
- Penguin
- Codfish
- Blue Whale
- Squid

Demonstration Supplies Listed By Chapter

Chapter	Supplies Needed
1: Observation Walk	No supplies needed
2: Cat's Eyes	Toilet paper tube, Foil, Rubber bands, Construction paper, Flashlight
3: Giraffe Saliva	Cornstarch, Water, Leaves and twigs, 2 Cups
4: Reptile Egg	Clear glass, Vinegar, Egg, Plastic wrap, Rubber band
5: Ear Cooling	Hot water, 1 Coffee cup, 1 8"x10" Pan, Instant thermometer
6: Making Butter	1 Pint of cream, 1 Large glass jar with lid, ½ Cup of water
7: Insect Hunt	A piece of fruit, Honey or syrup, Plate
8: Rainforest in a Bottle	2-Liter Soda bottle with top, Gravel, Potting soil, Several small plants, Scissors, Tape, Water
9: Examining Life Cycles	Life Cycle of a Butterfly Worksheet and Life Cycle of a Frog Worksheet (Appendix pp. 139-142) (Optional: Get a "Grow a Frog" or "Raise a Butterfly" kit ordered from a science supply store.)
10: Pouch Living	2 Thermometers, Large felt rectangle, Tape, Plastic baggie (sealable), Warm water
11: Bird Beaks	Chopsticks, Tweezers, Pliers, Eye dropper, Sugar water or honey, Gummy worms, Unshelled peanuts, Seeds, Raisins, Plate
12: Owl Pellet	Owl Pellet Dissection Kit (Order this from a science supply store.)
13: Primate Eyes	No supplies needed
14: Hairy Fur	2 Glass jars, Box at least 2 inches wider and taller than the jars, Cotton balls, 2 Thermometers
15: Blubber	1 Large plastic bag, Rubber band, Plastic glove, Shortening, Tub of ice water, Stopwatch
16: Fish Gills	Old Barbie or pony doll with hair, Tub of water
17: Echolocation	Shallow bowl, Water, Digital camera
18: Zoology Bingo	Zoology Bingo Cards (Download these for free from Elemental Science.)

STEAM Project Supplies Listed By Chapter

The multi-chapter and specific chapter STEAM projects listed in this guide are optional, so you may not need all of these supplies. However, this list has been provided for your convenience. If you do decide to do these projects, in addition to the items listed each week you will need glue, scissors, a variety of paint colors, and a set of markers.

Chapter	Supplies Needed
1	No supplies needed
2	Poster board or Shoebox, Paper plate, Pompom
3	Grey sock, Newspaper or cotton batting, Clothespin, Cardboard
4	Egg carton, Brown pipe clears, Googly eyes
5	Newspaper, Flour and salt, Pipe cleaners and cheesecloth
6	Air dry clay, Ingredients for cookies (white sugar, butter-flavored shortening, honey, eggs, vanilla extract, baking soda, all-purpose flour, and cinnamon)
7	Black pipe cleaners, Googly eyes, Styrofoam balls (one large, one small)
8	Toilet paper tube, Brown construction paper, Tissue paper (black, orange, yellow, white, brown, and green)
9	Paper (copy and construction)
10	Ingredients for eggs (mushrooms, oil, tomato, eggs, milk, and cream cheese), Fused beads, Empty can, Cotton balls, Felt (pink and blue)
11	Felt (white and black), Googly eyes, Paper (copy and construction)
12	Bird feeder kit or a soda bottle, Ingredients for deer cookies (Nutter Butter cookie, pretzel, M&M, and frosting)
13	Popsicle stick, Grey pipe cleaners, Black felt, Black beads
14	Decorative gourd, Leaves and twigs, Cloves
15	Chalk pastels, Blue construction paper, Ingredients for cheese (goat milk, vinegar, and salt), Cheesecloth, Thermometer, Pot
16	Egg carton, Construction paper, Paper clips, Magnet, Dowel rod, String
17	2 Gallon plastic jug, Googly eyes, Hot dogs
18	Blank map

THE SASSAFRAS GUIDE TO THE CHARACTERS FOUND IN VOLUME 1: ZOOLOGY

THROUGHOUT THE BOOK

* **Blaine Sassafras** – The male Sassafras twin, also known as Train.
* **Tracey Sassafras** – The female Sassafras twin, also known as Blaisey.
* **Uncle Cecil** – The Sassafras twins' crazy, but talented uncle.
* **President Lincoln** – Uncle Cecil's lab assistant, who also happens to be a prairie dog.
* **The Man With No Eyebrows** – He has no eyebrows and seems to be trying to sabotage the twins at every stop.

THE AFRICAN GRASSLAND (CHAPTERS 2-3)

* **Nicholas Mzuri** – (muh-zur-ee) The local expert for the African Grasslands leg of the twins' adventure. He is the owner and tour guide extraordinaire for Mzuri tours.
* **Hank** – A narcoleptic tourist who is a member of the twins' tour through the African Grasslands.
* **Pam and Fred** – The eccentric and hilarious couple who is a member of the twins' tour through the African Grasslands.
* **Shelley** – Nicholas Mzuri's assistant, who also happens to be a machete.
* **Imani Mzuri** – (ee-man-ee) Nicholas Mzuri's sister, who also helps out with Mzuri tours.

THE EGYPTIAN DESERT (CHAPTERS 4-5)

* **Princess Talibah** – (tal-ib-ah) The local expert for the Egyptian Desert leg of the twins' adventure. She is the princess of the Tuareg (twar-ehg) nomadic people.
* **Jendayi** – (jen-day-ee) Princess Talibah's hand-maiden and friend.
* **Hanif** – (han-if) Princess Talibah's advisor and teacher. He thinks that answers can be found in the stars, not in science.
* **Abasi** – (ah-bah-see) Princess Talibah's sworn protector. He is also in love with the princess.
* **Itja** – (eet-jah) The scoundrel leader of a group of bandits known as the Kekeway (kee-kee-way).
* **Mesneh** – (mez-nuh) He helps Princess Talibah teach Itja and his men a lesson after the Kekeway ravage his village.

THE CANADIAN FARM (CHAPTERS 6-7)

* **Jet (Jethro Mecklen, Jr.)** – The local expert for the Canadian Farm leg of the twins' adventure. He is also sixteen years old and friends with Edbert.
* **Edbert Snarfuffel** – The goofy sixteen year old boy who works with the Sassafras twins on the Smitty farm in Canada.
* **Farmer Smith** – He is the owner of Smitty Farms.
* **Ed Lumbia** – He is the farm foreman at Smitty Farms.
* **Tank & Billy** – They are the sons of Farmer Smith. They like to pull pranks and create problems for the twins and their companions.

THE AMAZON RAINFOREST (CHAPTERS 8-9)

* **Alvaro Manihuari** – (al-vah-ro mah-nee-ar-ee) the local expert for the Amazon Rainforest leg of the twins' adventure. He is also the owner of the Out on a Limb guesthouse.

* **Arrio** – (rr-ee-o) Alvaro's assistant and helper at the Out on a Limb guesthouse. He is a native Peruvian.
* **Skip, Gannon, and Gretchen** – They are three trekkers that are also staying at the Out on a Limb guesthouse.
* **Violetta Perez** – (vee-o-leh-tah peh-rezz) One of the Perez twins who is staying at the Out on a Limb guesthouse with their father. She makes friends with the Sassafras twins.
* **Vancho Perez** – (vahn-ch-o) One of the Perez twins who is staying at the Out on a Limb guesthouse with their father. He makes friends with the Sassafras twins.
* **Ernesto Perez** – He is the father of Violetta and Vancho. He is also president of ProLog.
* **Ortiz** – (or-tee-zz) He is the foreman for ProLog. He works under Mr. Perez.

THE AUSTRALIAN DECIDUOUS FOREST (CHAPTERS 10 & 12)

* **Willy Day** – The local expert for the Australian leg of the twins' adventure. He is an Australian filmmaker working on a project in the Brown Mountain Forest.
* **Ethel** – She works at the local diner near where Blaine arrives. She makes a great plate of "hot-maybe" eggs.
* **Ralphy Dingo** – He is one of the infamous Feuding Brown Mountain Hermits.
* **Matty Mingo** – He is one of the infamous Feuding Brown Mountain Hermits.

THE CHINESE BAMBOO FOREST (CHAPTERS 11 & 13)

* **Tashi Yidro** – (tah-see yee-dro) The local expert for the Chinese leg of the twin's adventure. She is a student at the local university who takes Tracey back to her home village.
* **Llamo** – (lahmo) She is Tashi's sister.
* **Norbu** – (nor-boo) He is Tashi's brother.
* **Amala** – (ah-mah-lah) She is Tashi, Llamo, and Norbu's mother.

THE ARCTIC (CHAPTERS 14-15)

* **Summer Beach** – The local expert for the Arctic leg of the twins' adventure. She was also a former classmate of Uncle Cecil's.
* **Ulysses S. Grant** – Summer Beach's lab assistant, who also happens to be an arctic ground squirrel.
* **Brooks Hirebro** – He is a professional snowboarder, entrepreneur, and also a friend of Summer's.
* **Yotimo** – (yo-tee-mo) He is an Inuit sled driver who finds Tracey in the arctic tundra; he is also a friend of Summer's.

THE ATLANTIC OCEAN (CHAPTERS 16-17)

* **Captain James Q. McScruffy** – The local expert for the Atlantic Ocean leg of the twins' adventure. He is a fisherman and the owner of the *Scot's Folly III*.
* **William Atwater** – He is the first mate on the *Scot's Folly III*.
* **Peach Beard** – The leader of the P.R.O. pirates, a band of men dedicated to bringing pirates back to the high seas.

CHAPTER LESSONS

CHAPTER 1: GRID SCHEDULE

Supplies Needed	
Demo	• No Supplies Needed
Projects	• No Additional Supplies Needed

Chapter Summary
Blaine and Tracey Sassafras begin this chapter upset because they are going to spend all summer with their crazy Uncle Cecil learning about science, instead of enjoying the zip lines at Camp Zip-fire. They are surprised to find out that Uncle Cecil and his lab assistant, President Lincoln, have a zip line adventure of sorts already prepared for them. The scientific duo has created invisible lines that will take the twins all over the world to learn about science. They will have to meet local experts along the way and enter data into their smartphones before they can proceed to the next place. At the close of the chapter, the twins take off for their first location, the African Grasslands.

Weekly Schedule				
	Day 1	**Day 2**	**Day 3**	**Day 4**
Read	☐ Read the section entitled "Crazy Uncle Cecil" of Chapter 1 in *SSA* Volume 1: Zoology*.	☐ (*Optional*) Read one or all of the assigned pages from the encyclopedia of your choice.	☐ Read the section entitled "Zip lines and Smartphones" of Chapter 1 in *SSA Volume 1: Zoology*.	☐ (*Optional*) Read one of the additional library books.
Write	☐ Set up the students' SCIDAT logbooks.	☐ Write observations learned from the demonstration on SL p. 5. ☐ (*Optional*) Write a narration on the Zoology Notes Sheet on SL p. 6.	☐ Go over the vocabulary word and enter it into the Zoology Glossary on SL** p. 93.	☐ (*Optional*) Complete the copywork or dictation assignment and add it to the Zoology Notes sheet on SL p. 6.
Do		☐ Do the demonstration entitled "Observation Walk."		☐ (*Optional*) Play a game of "I Spy."

*SSA = *The Sassafras Science Adventures*
**SL = *The Official Sassafras SCIDAT Logbook: Zoology Edition*

CHAPTER 1: LIST SCHEDULE

CHAPTER SUMMARY

Blaine and Tracey Sassafras begin this chapter upset because they are going to spend all summer with their crazy Uncle Cecil learning about science, instead of enjoying the zip lines at Camp Zip-fire. They are surprised to find out that Uncle Cecil and his lab assistant, President Lincoln, have a zip line adventure of sorts already prepared for them. The scientific duo has created invisible lines that will take the twins all over the world to learn about science. They will have to meet local experts along the way and enter data into their smartphones before they can proceed to the next place. At the close of the chapter, the twins take off for their first location, the African Grasslands.

ESSENTIALS

Read

- ☐ Read the section entitled "Crazy Uncle Cecil" of Chapter 1 in *SSA* Volume 1: Zoology*.
- ☐ Read the section entitled "Zip lines and Smartphones" of Chapter 1 in *SSA Volume 1: Zoology*.

Write

- ☐ Set up the students' SCIDAT logbooks.
- ☐ Go over the vocabulary word and enter it into the Zoology Glossary on SL** p. 93.
- ☐ Write observations learned from the demonstration on SL p. 5.

Do

- ☐ Do the demonstration entitled "Observing Walk."

(OPTIONAL) EXTRAS

Read

- ☐ Read one of the additional library books.

Write

- ☐ Write a narration on the Zoology Notes Sheet on SL p. 6.
- ☐ Complete the copywork or dictation assignment and add it to the Zoology Notes sheet on SL p. 6.

Do

- ☐ Play a game of "I Spy."

Supplies Needed	
Demo	• No Supplies Needed
Projects	• No Additional Supplies Needed

*SSA = *The Sassafras Science Adventures*
**SL = *The Official Sassafras SCIDAT Logbook: Zoology Edition*

CHAPTER 1: THE ADVENTURE BEGINS

READ: GATHERING INFORMATION

LIVING BOOK READING ASSIGNMENT

- 📖 Chapter 1 of *The Sassafras Science Adventures Volume 1: Zoology*

(OPTIONAL) ENCYCLOPEDIA READINGS

- 🔎 *Kingfisher First Encyclopedia of Animals* – No pages scheduled for this week.
- 🔎 *DK Encyclopedia of Animals* pp. 14-15 (Animal Classification), pp. 16-17 (Animal Kingdoms)

(OPTIONAL) ADDITIONAL LIBRARY BOOKS

- 📖 *What Is the Animal Kingdom?* (Science of Living Things) by Bobbie Kalman
- 📖 *Animal Classification* by Polly Goodman
- 📖 *Who's in Your Class?* Level 4: An Animal Adventure (Lithgow Palooza Readers: Level 4) by John Lithgow and Susan Blackaby

WRITE: KEEPING A NOTEBOOK

SCIDAT LOGBOOK SHEETS

This week, you will set up the students' SCIDAT logbook. You can use blank sheets of copy paper with dividers for each section or purchase *The Official Sassafras SCIDAT Logbook: Zoology Edition* with all the pictures from Elemental Science. For each of these sheets you can have the students enter information only from *The Sassafras Science Adventures Volume 1: Zoology*, or you can have them do additional research to gather more facts. The following video shares a peek inside a 2nd-grader's SCIDAT Logbook:

🖱 https://www.youtube.com/watch?v=0m4nj-K7s58

What you choose to do will depend upon the ages and abilities of your students. Below is an explanation of each of the student sheets.

Habitat Information Sheet

The purpose of these sheets is for the students to record what they have learned about the various habitats visited in *The Sassafras Science Adventures Volume 1: Zoology*. (NOTE—The chapters should have most of the information for the sheet. If not, you can choose to leave those sections blank or use the encyclopedia readings to fill in the answers.)

HABITAT: Enter the name of the habitat they are studying.

LOCATION AND LOCAL EXPERT: Enter the place that the particular habitat was found along with the name of the local expert from the region.

AVERAGE RAINFALL: Enter the average rainfall the habitat receives per year.

AVERAGE TEMPERATURE: Enter information about the average temperatures throughout the year in the habitat, e.g. "warm summers and cold winters".

MAIN CHARACTERISTICS: Enter the main features of the habitat, e.g. "lots of grass, very few trees".

ANIMALS FOUND THERE: Enter the animals from the story that are found in the habitat. You can write the names of the animals or use pictures of them.

Around the World

The purpose of these sheets is to give the students an opportunity to work on their mapping skills. (NOTE—The information for this sheet will come from the encyclopedia readings.)

MAP: Color the places on the world map where the particular habitat can be found.

CONTINENTS FOUND ON: Enter the continents where the particular habitat can be found around the globe.

Animal Record Sheets

The purpose of these sheets is for the students to record what they have learned about the various animals that are introduced in *The Sassafras Science Adventures Volume 1: Zoology*.

ANIMAL NAME: Enter the name of the animal.

CLASSIFICATION: Enter whether it is a mammal, bird, reptile, amphibian, fish or invertebrate.

FOOD: Enter whether the animal is a carnivore, herbivore or omnivore.

LOCATION FOUND: Enter the habitat it was found in.

INFORMATION LEARNED: Enter any information that the students learned about the animal.

Zoology Notes Sheets

The purpose of these sheets is for the students to record any additional information that they have learned during their study of zoology. You can use these sheets to record narrations, copywork, and dictation assignments.

Project Record Sheets

The purpose of these sheets is for the students to record the STEAM projects they have done through the course of their study of zoology.

Zoology Glossary

The purpose of the glossary is for the students to create a glossary of terms that they have encountered throughout reading *The Sassafras Science Adventures Volume 1: Zoology*. They can look each of the terms up in a science encyclopedia or in the glossary included on pp. 145-146 of this guide. The students should illustrate each of the vocabulary words on their own. (NOTE—In *The Official Sassafras SCIDAT Logbook: Zoology Edition* these pictures are already provided.)

VOCABULARY

Have the older students look up the following terms in the glossary in the appendix on pp. 145-146 or in a science encyclopedia. Then, have them copy the definition onto a blank index card or into their SCIDAT logbook.

☞ CLASSIFICATION – A way of identifying or grouping living things.

☞ OBSERVATION – Something that you see with your eyes, or the act of regarding attentively.

(OPTIONAL) COPYWORK

Copywork Selection

Observation is taking the time to look at the things around me.

Dictation Passage (Poem selection by Henry Wadsworth Longfellow)

And he wandered away and away,
With Nature the dear old nurse,
Who sang to him night and day,
The rhymes of the universe.
And when the way seemed long,
and his heart began to fail,
She sang a more wonderful song,
or told a more wonderful tale.

DO: PLAYING WITH SCIENCE

SCIENTIFIC DEMONSTRATION: OBSERVATION WALK

Begin by taking a moment to discuss what nature study is and the importance of observation in science. You can view the following blog posts for more information on the subject:

🖱 https://elementalscience.com/blogs/podcast/episode-8

🖱 http://elementalscience.com/blogs/news/63858627-observation-is-key

Explain that today you are going to practice your observation skills while on a walk. Then, take a walk in your neighborhood or on a nature trail nearby where you live. Allow the students to make observations and ask questions. Ask the students:

? What kind of plants do you see?

? What kind of animals do you see?

? What else do you see that you would like to tell me about?

(OPTIONAL) STEAM PROJECTS

Activities For This Chapter

✂ I SPY – Play a game of "I Spy" to help the students to work on their observation skills.

CHAPTER 2: GRID SCHEDULE

Supplies Needed	
Demo	• Toilet paper tube, Foil, Rubber bands, Construction paper, Flashlight
Projects	• Poster board or Shoebox, Paper plate, Pompom

Chapter Summary
The chapter opens with Blaine and Tracey arriving at their first stop in the African grasslands. They join Nicholas Mzuri and four other guests on a photo safari through Kenya. The group spots a pride of lions and Nicholas shares more about them. They also race several cheetahs on the hunt and their guide tells more about the animals. All the while, the narcoleptic Hank will fall asleep and the comical couple, Fred and Pam, will make their characteristic blunders and mistakes. The chapter ends with the mysterious Man With No Eyebrows stealing the jeep, leaving the group stranded in Elephant Valley.

Weekly Schedule

	Day 1	Day 2	Day 3	Day 4
Read	☐ Read the section entitled "Look... a Lion" of Chapter 2 in *SSA Volume 1: Zoology.*	☐ Read the section entitled "Chasing Cheetahs" of Chapter 2 in *SSA Volume 1: Zoology.*	☐ (*Optional*) Read one or all of the assigned pages from the encyclopedia of your choice.	☐ (*Optional*) Read one of the additional library books.
Write	☐ Fill out the Animal Record Sheet on SL p. 9 for the lion. ☐ Go over the vocabulary words and enter them into the Zoology Glossary on SL pp. 93-94.	☐ Fill out the Animal Record Sheet on SL p. 10 for the cheetah. ☐ Add to the Habitat Information Sheet on SL p. 7 for the African Grasslands.	☐ Write the information learned from the demonstration on SL p. 13. ☐ (*Optional*) Write a narration on the Zoology Notes Sheet on SL p. 13.	☐ (*Optional*) Do the copywork or dictation assignment and add it to the Zoology Notes sheet on SL p. 13.
Do	☐ (*Optional*) Make a Handprint Lion.	☐ (*Optional*) Make a Cheetah Mask. ☐ (*Optional*) Have a Foot Race.	☐ Do the demonstration entitled "Cat's Eyes."	☐ (*Optional*) Add the animals studied this week to the food chart and habitat posters.

CHAPTER 2: LIST SCHEDULE

CHAPTER SUMMARY

The chapter opens with Blaine and Tracey arriving at their first stop in the African grasslands. They join Nicholas Mzuri and four other guests on a photo safari through Kenya. The group spots a pride of lions and Nicholas shares more about them. They also race several cheetahs on the hunt and their guide tells more about the animals. All the while, the narcoleptic Hank will fall asleep and the comical couple, Fred and Pam, will make their characteristic blunders and mistakes. The chapter ends with the mysterious Man With No Eyebrows stealing the jeep, leaving the group stranded in Elephant Valley.

ESSENTIALS

Read

☐ Read the section entitled "Look... a Lion" of Chapter 2 in *SSA Volume 1: Zoology.*

☐ Read the section entitled "Chasing Cheetahs" of Chapter 2 in *SSA Volume 1: Zoology.*

Write

☐ Fill out the Animal Record Sheet on SL p. 9 for the lion.

☐ Go over the vocabulary words and enter them into the Zoology Glossary on SL pp. 93-94.

☐ Fill out the Animal Record Sheet on SL p. 10 for the cheetah.

☐ Add to the Habitat Information Sheet on SL p. 7 for the African Grasslands.

☐ Write the information learned from the demonstration on SL p. 13.

Do

☐ Do the demonstration entitled "Cat's Eyes."

(OPTIONAL) EXTRAS

Read

☐ Read one of the additional library books.

☐ Read one or all of the assigned pages from the encyclopedia of your choice.

Write

☐ Write a narration on the Zoology Notes Sheet on SL p. 13.

☐ Complete the copywork or dictation assignment and add it to the Zoology Notes sheet on SL p. 13.

Do

☐ Make a Handprint Lion.

☐ Make a Cheetah Mask or have a Foot Race.

☐ Add the animals studied this week to the food chart and habitat posters.

Supplies Needed	
Demo	• Toilet paper tube, Foil, Rubber bands, Construction paper, Flashlight
Projects	• Poster board or Shoebox, Paper plate, Pompom

CHAPTER 2: THE AFRICAN GRASSLANDS

READ: GATHERING INFORMATION

LIVING BOOK READING ASSIGNMENT

📖 Chapter 2 of *The Sassafras Science Adventures Volume 1: Zoology*

(OPTIONAL) ENCYCLOPEDIA READINGS

🔎 *Kingfisher First Encyclopedia of Animals* p. 20 (Mammals), p. 21 (Lion), p. 24 (Cheetah)

🔎 *DK Encyclopedia of Animals* pp. 239-241 (Mammals), pp. 231-233 (Lion), pp. 136-137 (Cheetah)

(OPTIONAL) ADDITIONAL LIBRARY BOOKS

📖 *Face to Face with Lions* (Face to Face with Animals) by Dereck Joubert and Beverly Joubert

📖 *Tawny Scrawny Lion* (Little Golden Book) by Golden Books and Gustaf Tenggren

📖 *The Cheetah: Fast as Lightning* (Animal Close-Ups) by Christine Denis-Huot and Michel Denis-Huot

📖 *Cheetah* (Welcome Books: Animals of the World) by Edana Eckart

📖 *Cheetah Cubs: Station Stop 2* (All Aboard Science Reader) by Ginjer L. Clarke and Lucia Washburn

📖 *What is a Mammal?* (Science of Living Things) by Kalman

WRITE: KEEPING A NOTEBOOK

SCIDAT LOGBOOK SHEETS

This week, you can have the students begin to fill out a Habitat Information Sheet for the African Grasslands and a logbook page for lions and cheetahs. The students could use the following information:

Habitat Information Sheet - Grassland

HABITAT: Grasslands (Tropical)

LOCATION AND LOCAL EXPERT: Africa (or Kenya), and Nicholas Mzuri

AVERAGE RAINFALL: The grasslands receive an average of 10 to 30 inches per year.

AVERAGE TEMPERATURE: The grasslands usually have warm winters and hot summers.

MAIN CHARACTERISTICS: There are few trees, vast grassy fields, and some rolling hills.

ANIMALS FOUND THERE: Add the cheetah and lion (The students could also add zebra and wildebeest).

Animal Record Sheet - Lion

ANIMAL NAME: Lion

CLASSIFICATION: Mammal

FOOD: Carnivore

LOCATION FOUND: African Grasslands

INFORMATION LEARNED

- They are part of the cat family.
- They are found in the plains of Africa, but they once roamed throughout Europe, Africa and Asia.
- They are the grassland's largest predator.
- They are social animals that live in groups called prides, which are made up of several females and their cubs along with a few males-one of whom is dominant.
- They hunt at night and rest during the day.
- They hunt anything that they can kill, but usually they eat zebra and wildebeest.
- Lion cubs have spots.
- Male lions have large manes around their necks.
- Lions lie down to eat and rest with their paws in front.

Animal Record Sheet - Cheetah

ANIMAL NAME: Cheetah

CLASSIFICATION: Mammal

FOOD: Carnivore

LOCATION FOUND: African Grasslands

INFORMATION LEARNED

- They have slim, muscular bodies with long legs that allow them to move with speed, strength and flexibility.
- They use their tails to help them balance while leaping, running, and climbing.
- They can reach speeds of over sixty miles per hour, but only for short distances.
- They are found in the plains of Africa, south of the Sahara, but once they lived in North Africa, the Middle East and India.
- Female cheetahs can have up to four babies at a time.
- Cheetah babies are born with a coat of long gray hair so that they look more like a honey badger, which is an aggressive animal. This is a form of camouflage.
- They are endangered animals.

VOCABULARY

Have the older students look up the following terms in the glossary in the appendix on pp. 145-146 or in a science encyclopedia. Then, have them copy the definitions onto a blank index card or into their SCIDAT logbook.

- ⟁ FOOD CHAIN – A chain of living things that eat each other.
- ⟁ GRASSLAND – A habitat characterized by vast grassy fields.

◌ MAMMALS – Any warm-blooded vertebrate with skin that is more or less covered with hair; they give birth to live young which are nourished with milk at the beginning of their life.

(OPTIONAL) COPYWORK

Copywork Sentence

Mammals are warm-blooded. They feed their young with milk.

Dictation Selection

Mammals are warm-blooded and they feed their young with milk. Mammals also have fur or hair covering most of their body. Lions and cheetahs are mammals.

DO: PLAYING WITH SCIENCE

SCIENTIFIC DEMONSTRATION: CAT'S EYES

Materials
- ☑ Toilet paper tube
- ☑ Foil
- ☑ Rubber bands
- ☑ Construction paper
- ☑ Flashlight

Procedure
1. Cover one end of one of the toilet paper tubes with the foil and cover the other with construction paper. Secure both with rubber bands.
2. Take your tubes and your flashlight and go into a dark room.
3. Shine the flashlight into the tubes and observe what happens.

Explanation

The students should see that the tube with the foil on the bottom appears to shine. This is because the foil reflects light, just like a cat's eye. They reflect light which makes them appear to glow in the dark.

Take It Further

Have the students see what happens to their pupils in the presence of light by going into a dark room and turning on the lights. Have them watch each others' eyes to see what happens.

(OPTIONAL) STEAM PROJECTS

Multi-chapter Activities

✂ FOOD CHART – Create a poster with three sections on it—one for carnivores, one for omnivores and one for herbivores. You will add to this weekly over the coming weeks, so post it somewhere visible on your wall. This week, you can add the lion and cheetah to the carnivore side of your food chart. You can use the mini-animal pictures that are in the appendix on p. 129 of this guide.

✂ HABITAT PROJECT – Make a poster or diorama that depicts the grassland habitat. This week, add the lion and cheetah. You can use the mini-animal pictures found in the appendix on p. 129 of this guide and the habitat poster on p. 130 or print out your own.

Activities For This Chapter

✂ MAKE A HANDPRINT LION – This project is super cute! It uses the students hands and some orange and yellow paint to make a lion design. Simply follow the model project found on this pin:

🖱 https://www.pinterest.com/pin/192036371585041320/

✂ MAKE A CHEETAH MASK – Paint a paper plate completely yellow. Then use a pompom dipped in brown paint to add spots to the plate. Cut out holes for the eyes and add two yellow triangles for ears. Finally add a string so that the students can wear their masks.

✂ HAVE A FOOT RACE – Have a foot race and declare the winner to be the cheetah of the group! Crown them with the cheetah mask you made earlier.

CHAPTER 3: GRID SCHEDULE

Supplies Needed	
Demo	• Cornstarch, Water, Leaves, Twigs, 2 Cups
Projects	• Grey sock, Newspaper or cotton batting, Clothespin, Cardboard

Chapter Summary

The chapter opens with Blaine, Tracey, Hank, Fred, Pam, and Nicholas, stuck in Elephant Valley with no vehicle. Uncle Cecil tells the twins that the only way out is to continue to gather and enter their scientific data. So, the twins stay with the group and learn from Nicholas about the elephants that live in the valley. As night falls, they build a shelter out of acacia trees to sleep in. Early the next morning, they are awakened by giraffes eating their shelter and the twins get a chance to learn all about giraffes from their guide. They decide to trek back to the lodge. The chapter ends with a herd of stampeding zebras that threatened to run the group over.

Weekly Schedule

	Day 1	Day 2	Day 3	Day 4
Read	☐ Read the section entitled "Eyeing Elephants" of Chapter 3 in *SSA Volume 1: Zoology.*	☐ Read the section entitled "Glimpsing Giraffes" of Chapter 3 in *SSA Volume 1: Zoology.*	☐ (*Optional*) Read one or all of the assigned pages from the encyclopedia of your choice.	☐ (*Optional*) Read one of the additional library books.
Write	☐ Fill out the Animal Record Sheet on SL p. 11 for the elephant. ☐ Go over the vocabulary words and enter them into the Zoology Glossary on SL p. 94.	☐ Fill out the Animal Record Sheet on SL p. 12 for the giraffe. ☐ Add to the Habitat Information Sheet on SL p. 7 for the African Grasslands.	☐ Write the information learned from the demonstration on SL p. 14. ☐ (*Optional*) Write a narration on the Zoology Notes Sheet on SL p. 14.	☐ (*Optional*) Complete the copywork or dictation assignment and add it to the Zoology Notes sheet on SL p. 14. ☐ (*Optional*) Take Zoology Quiz #1.
Do	☐ (*Optional*) Pretend to be an elephant. ☐ (*Optional*) Make a Elephant Trunk.	☐ (*Optional*) Make a Giraffe.	☐ Do the demonstration entitled "Giraffe Saliva."	☐ (*Optional*) Add the animals studied this week to the food chart and habitat posters.

CHAPTER 3 : LIST SCHEDULE

CHAPTER SUMMARY

The chapter opens with Blaine, Tracey, Hank, Fred, Pam, and Nicholas, stuck in Elephant Valley with no vehicle. Uncle Cecil tells the twins that the only way out is to continue to gather and enter their scientific data. So, the twins stay with the group and learn from Nicholas about the elephants that live in the valley. As night falls, they build a shelter out of acacia trees to sleep in. Early the next morning, they are awakened by giraffes eating their shelter and the twins get a chance to learn all about giraffes from their guide. They decide to trek back to the lodge. The chapter ends with a herd of stampeding zebras that threatened to run the group over.

ESSENTIALS

Read

☐ Read the section entitled "Eyeing Elephants" of Chapter 3 in *SSA Volume 1: Zoology*.

☐ Read the section entitled "Glimpsing Giraffes" of Chapter 3 in *SSA Volume 1: Zoology*.

Write

☐ Fill out the Animal Record Sheet on SL p. 11 for the elephant.

☐ Go over the vocabulary words and enter them into the Zoology Glossary on SL p. 94.

☐ Fill out the Animal Record Sheet on SL p. 12 for the giraffe.

☐ Add to the Habitat Information Sheet on SL p. 7 for the African Grasslands.

☐ Write the information learned from the demonstration on SL p. 14.

Do

☐ Do the demonstration entitled "Giraffe Saliva."

(OPTIONAL) EXTRAS

Read

☐ Read one of the additional library books.

☐ Read one or all of the assigned pages from the encyclopedia of your choice.

Write

☐ Write a narration on the Zoology Notes Sheet on SL p. 14.

☐ Complete the copywork or dictation assignment and add it to the Zoology Notes sheet on SL p. 14.

☐ Take Zoology Quiz #1.

Do

☐ Pretend to be an elephant and make a Elephant Trunk.

☐ Make a Giraffe.

☐ Add the animals studied this week to the food chart and habitat posters.

Supplies Needed	
Demo	• Cornstarch, Water, Leaves, Twigs, 2 Cups
Projects	• Grey sock, Newspaper or cotton batting, Clothespin, Cardboard

CHAPTER 3: LOST IN ELEPHANT VALLEY

READ: GATHERING INFORMATION

LIVING BOOK READING ASSIGNMENT

📖 Chapter 3 of *The Sassafras Science Adventures Volume 1: Zoology*

(OPTIONAL) ENCYCLOPEDIA READINGS

🔍 *Kingfisher First Encyclopedia of Animals* p. 26 (Elephant), p. 34 (Giraffe)

🔍 *DK Encyclopedia of Animals* pp. 170-172 (Elephant), pp. 187-189 (Giraffe), pp. 64-65 (Grasslands)

(OPTIONAL) ADDITIONAL LIBRARY BOOKS

📖 *Elephants: A Book for Children* by Steve Bloom

📖 *Face to Face With Elephants* (Face to Face with Animals) by Beverly Joubert

📖 *Giraffes* by Jill Anderson

📖 *Baby Giraffes* (It's Fun to Learn about Baby Animals) by Bobbie Kalman

📖 *Chee-Lin: A Giraffe's Journey* by James Rumford

WRITE: KEEPING A NOTEBOOK

SCIDAT LOGBOOK SHEETS

This week, you can have the students fill out a logbook page for the elephant and the giraffe, as well as continue to fill out a Habitat Information Sheet and the Around the World Sheet for the African Grasslands. The students could use the following information:

Habitat Information Sheet - Grasslands

HABITAT: Grasslands (Tropical)

MAIN CHARACTERISTICS: The grasslands have the occasional watering hole and few rocks.

ANIMALS FOUND THERE: Add the elephant and giraffe (The students could also add zebra).

Around the World - Grassland Habitat

MAP: Have the students put a star in the middle of Kenya and label it "Kenya". Then, have them color the regions of the world that are covered with the tropical grasslands habitat light green. Use the map pictured as a guide.

CONTINENTS FOUND ON: North America, South America, Africa, Asia, Australia or Oceania

Animal Record Sheet - Elephant

ANIMAL NAME: Elephant

CLASSIFICATION: Mammal

FOOD: Herbivore

LOCATION FOUND: African Grasslands

INFORMATION LEARNED

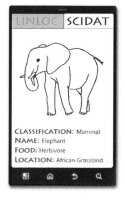

- There are two different species, one found on the African grasslands and another in the forests of Asia.
- They are very intelligent and have good memories.
- They travel in family units of eight to ten elephants, called herds.
- They are the heaviest land animals.
- The herds are usually composed of one dominant male and related females. Single males can live on their own or in a bachelor herd.
- They can use their trunks as an arm for eating, drinking and carrying or moving heavy loads.
- They also use their trunks to make a trumpeting call to warn other animals and communicate with the herd.
- They have large grinding cheek teeth that wear down and are replaced from the rear.
- Their tusks are extended upper teeth that they use for defense and for digging for food and water.
- Their ears are very large and are used to keep the animal cool as well as in communication and defense.

Animal Record Sheet - Giraffe

ANIMAL NAME: Giraffe

CLASSIFICATION: Mammal

FOOD: Herbivore

LOCATION FOUND: African Grasslands

INFORMATION LEARNED

- They are the world's tallest animal, up to 20 feet in height.
- Their long necks contain just 7 vertebrae, just like other mammals, which are greatly elongated.
- They use their height to graze from the tops of trees, especially the acacia tree.
- They eat leaves, twigs, shoots, and other vegetation.
- Female giraffes tend to feed on lower trees and shrubs, while males stretch to eat from the top.
- They use their long purple tongue and grooved canine teeth to strip the leaves and their thick lips protect them from the thorns of the acacia tree.
- The eight sub species of giraffe can be differentiated by differences in their coat's color and pattern type, but each individual giraffe has a unique pattern, just like a fingerprint.
- Their spots, that can be big or blotchy, and the spotted pattern helps with camouflage.
- Both male and female have short horns that don't shed each year, but males use them when fighting.
- Female giraffes give birth after fifteen months, and the calf can walk an hour or two after birth.
- Their tail is tipped with long hairs for swatting away insects.
- Giraffes can canter and gallop on their two-toed hoofed feet.

VOCABULARY

Have the older students look up the following terms in the glossary in the appendix on pp. 145-146 or in a science encyclopedia. Then, have them copy the definitions onto a blank index card or into their SCIDAT logbook.

- ✏ HERBIVORE – An animal that eats plants.
- ✏ CARNIVORE – An animal that eats meat.
- ✏ OMNIVORE – An animal that eats both plants and meat.

(OPTIONAL) COPYWORK

Copywork Sentence

Herbivores are animals that eat only plants.

Dictation Selection

Herbivores are animals whose diet consists of only plants. Carnivores are animals whose diet consists of only meat. Omnivores are animals whose diet consists of both plants and meat. Elephants and giraffes are herbivores, while lions and cheetahs are carnivores.

(OPTIONAL) QUIZ

This week, you can give the students a quiz based on what they learned in chapters 2 and 3. You can find the quiz in the appendix on p. 155.

Quiz #1 Answers
1. D
2. B
3. C
4. D
5. C

DO: PLAYING WITH SCIENCE

SCIENTIFIC DEMONSTRATION: GIRAFFE SALIVA

Materials
- ☑ Cornstarch
- ☑ Water
- ☑ Leaves and twigs
- ☑ 2 Cups

Procedure
1. Mix 1 cup of cornstarch and 1/3 cup of water together in one glass. This is your giraffe saliva.
2. In another glass, add one cup of water. This is your human saliva.
3. Mix in several leaves and a few small twigs to both glasses.
4. Observe the two cups and ask the students:

? In which is it easier to see the twigs and leaves?

5. Gently press one of your fingers into each of the liquids and then ask the students:

? In which one can you feel the leaves and twigs?

Explanation

The saliva in a giraffe's mouth is much thicker, which protects the inside of its mouth from the thorns of the acacia tree from which it prefers to eat from.

Take It Further

Have the students try out other saliva ideas, such as soda water, milk, or slime to see how those would protect the mouth from twigs and leaves.

(OPTIONAL) STEAM PROJECTS

Multi-chapter Activities

✂ FOOD CHART – This week, add the elephant and the giraffe to the herbivore side of your food chart. You can use the mini-animal pictures found in the appendix on p.129 of this guide or print out your own.

✂ HABITAT PROJECT – This week, add the elephant and the giraffe. You can use the mini-animal pictures found in the appendix on p. 129 of this guide or print out your own.

Activities For This Chapter

✂ ANIMAL NOISES – Pretend to be an elephant!

✂ MAKE YOUR OWN ELEPHANT TRUNK – Have the students stuff a gray sock with paper or cotton to give it more stability. Then staple an elastic band on either side of the sock, so that the sock will fit around their head and sit in front of their nose. Then have them pretend to be an elephant.

✂ MAKE A GIRAFFE – Make a clothespin giraffe by following the directions from this website:

🖰 http://www.busybeekidscrafts.com/Clothes-Pin-Giraffe.html

CHAPTER 4: GRID SCHEDULE

Supplies Needed	
Demo	• Clear glass, Vinegar, Egg, Plastic wrap, Rubber band
Projects	• Egg carton, Brown pipe clears, Googly eyes

Chapter Summary

The chapter opens with Blaine, Tracey, Hank, Fred, Pam, and Nicholas, being rescued from the zebras by Imani, Nicholas's sister. She takes them back to the lodge where the twins say goodbye before leaving to Egypt that night. They land on a camel in the desert, in the dark. They listen as Princess Talibah shares about camels and then they are lulled to sleep by the motion of riding on a camel. When the twins wake up, they realize that they are being falsely accused of being camel thieves along with their local expert, Princess Talibah and her friends, Jendayi, Hanif, and Abasi. The group quickly realizes that they are being accused by a group of bandits, called the Kekeway. Blaine, Tracey, and the others are marched through the desert to an ancient tomb where they are left to wait for a surprise. The surprise happens to be a disappearing floor with a pit of cobras below them. The chapter ends with the group desperately trying to find a way out as Talibah shares more about the cobras.

Weekly Schedule

	Day 1	Day 2	Day 3	Day 4
Read	☐ Read the section entitled "Camel Criminals" of Chapter 4 in *SSA Volume 1: Zoology*.	☐ Read the section entitled "Captured with Cobras" of Chapter 4 in *SSA Volume 1: Zoology*.	☐ (*Optional*) Read one or all of the assigned pages from the encyclopedia of your choice.	☐ (*Optional*) Read one of the additional library books.
Write	☐ Fill out the Animal Record Sheet on SL p. 19 for the camel. ☐ Go over the vocabulary words and enter them into the Zoology Glossary on SL p. 95.	☐ Fill out the Animal Record Sheet on SL p. 20 for the cobra. ☐ Add to the Habitat Information Sheet on SL p. 17 for the Desert.	☐ Write the information learned from the demonstration on SL p. 23. ☐ (*Optional*) Write a narration on the Zoology Notes Sheet on SL p. 23.	☐ (*Optional*) Complete the copywork or dictation assignment and add it to the Zoology Notes sheet on SL p. 23.
Do	☐ Make an Egg Carton Camel.		☐ Do the demonstration entitled "Reptile Egg."	☐ (*Optional*) Add the animals studied this week to the food chart and habitat posters.

CHAPTER 4: LIST SCHEDULE

CHAPTER SUMMARY

The chapter opens with Blaine, Tracey, Hank, Fred, Pam, and Nicholas, being rescued from the zebras by Imani, Nicholas's sister. She takes them back to the lodge where the twins say goodbye before leaving to Egypt that night. They land on a camel in the desert, in the dark. They listen as Princess Talibah shares about camels and then they are lulled to sleep by the motion of riding on a camel. When the twins wake up, they realize that they are being falsely accused of being camel thieves along with their local expert, Princess Talibah and her friends, Jendayi, Hanif, and Abasi. The group quickly realizes that they are being accused by a group of bandits, called the Kekeway. Blaine, Tracey, and the others are marched through the desert to an ancient tomb where they are left to wait for a surprise. The surprise happens to be a disappearing floor with a pit of cobras below them. The chapter ends with the group desperately trying to find a way out as Talibah shares more about the cobras.

ESSENTIALS

Read

- ☐ Read the section entitled "Camel Criminals" of Chapter 4 in *SSA Volume 1: Zoology*.
- ☐ Read the section entitled "Captured with Cobras" of Chapter 4 in *SSA Volume 1: Zoology*.

Write

- ☐ Fill out the Animal Record Sheet on SL p. 19 for the camel.
- ☐ Go over the vocabulary words and enter them into the Zoology Glossary on SL p. 95.
- ☐ Fill out the Animal Record Sheet on SL p. 20 for the cobra.
- ☐ Add to the Habitat Information Sheet on SL p. 17 for the Desert.
- ☐ Write the information learned from the demonstration on SL p. 23.

Do

- ☐ Do the demonstration entitled "Reptile Egg."

(OPTIONAL) EXTRAS

Read

- ☐ Read one of the additional library books.
- ☐ Read one or all of the assigned pages from the encyclopedia of your choice.

Write

- ☐ Write a narration on the Zoology Notes Sheet on SL p. 23.
- ☐ Complete the copywork or dictation assignment and add it to the Zoology Notes sheet on SL p. 23.

Do

- ☐ Make an Egg Carton Camel.
- ☐ Add the animals studied this week to the food chart and habitat posters.

Supplies Needed	
Demo	• Clear glass, Vinegar, Egg, Plastic wrap, Rubber band
Projects	• Egg carton, Brown pipe clears, Googly eyes

CHAPTER 4: OFF TO THE DESERT

READ: GATHERING INFORMATION

LIVING BOOK READING ASSIGNMENT

- 📖 Chapter 4 of *The Sassafras Science Adventures Volume 1: Zoology*

(OPTIONAL) ENCYCLOPEDIA READINGS

- ☌ *Kingfisher First Encyclopedia of Animals* p. 35 (Camel), p. 110 (Cobra), p. 106 (Reptile)
- ☌ *DK Encyclopedia of Animals* pp. 124-126 (Camel), p. 143 (Cobra), pp. 297-299 (Reptile)

(OPTIONAL) ADDITIONAL LIBRARY BOOKS

- 📖 *Camels* (Nature Watch) by Cherie Winner
- 📖 *I Wonder Why Camels Have Humps: And Other Questions About Animals* by Anita Ganeri
- 📖 *King Cobras* by Joanne Mattern and Gail Saunders-Smith
- 📖 *Egyptian Cobra* (Killer Snakes) by Jessica O'Donnell
- 📖 *Smart Kids Reptiles* by Simon Mugford

WRITE: KEEPING A NOTEBOOK

SCIDAT LOGBOOK SHEETS

This week, you can have the students fill out a logbook page for the camel and the cobra, as well as begin to fill out a Habitat Information Sheet for the Egyptian Desert. The students could use the following information:

Habitat Information Sheet - Desert

HABITAT: Desert

LOCATION AND LOCAL EXPERT: Egypt and Princess Talibah

AVERAGE RAINFALL

AVERAGE TEMPERATURE

MAIN CHARACTERISTICS: The desert has lots of sand and dunes.

ANIMALS FOUND THERE: Add the camel and cobra.

Animal Record Sheet - Camel

ANIMAL NAME: Camel

CLASSIFICATION: Mammal

FOOD: Herbivore

LOCATION FOUND: Egyptian Desert

INFORMATION LEARNED

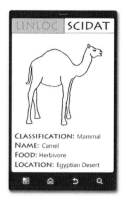

- They have humps of stored fat on their back to help them survive for days without food or water.
- They can endure many months without drinking water, as their digestive systems are extremely effective at extracting water from plants.
- They can drink up to thirty gallons in minutes to make up for the loss.
- They are the largest even toed mammal, with feet that are big and wide to prevent them from sinking into the sand.
- They have two rows of eyelashes and nostrils that can completely close, that prevent sand from getting into their eyes and nose.
- They live in the world's driest deserts.

Animal Record Sheet - Cobra

ANIMAL NAME: Cobra

CLASSIFICATION: Reptile

FOOD: Carnivore

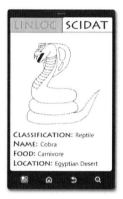

LOCATION FOUND: Egyptian Desert

INFORMATION LEARNED

- Cobras all have venom.
- Their venom stops the heart and lungs from working.
- Cobras inject their venom through their hollow teeth and it is produced in glands at the roof of the snake's mouth.
- When they feel threatened, cobras can flatten their necks to make themselves look bigger.
- Cobras are reptiles that have no legs, eyelids or external ears.
- They shed their skin up to six times a year as they grow. The new skin grows under the old and it splits in a process called sloughing, leaving behind a transparent version of the snake.
- They bury their eggs in a damp, warm place where they will hatch by themselves.
- Cobras eat small vertebrates.
- They kill them with their venom and then swallow them whole.
- Their jaws are elastic and able to stretch very wide.

VOCABULARY

Have your older students look up the following term in the glossary in the appendix on pp. 145-146 or in a science encyclopedia. Have them copy the definition onto a blank index card or into their SCIDAT logbook.

✐ REPTILE – A group of cold-blooded animals that usually have rough skin.

(OPTIONAL) COPYWORK

Copywork Sentence

Reptiles are cold-blooded animals with rough skin.

Dictation Selection

Reptiles are cold-blooded animals with scaly, watertight skin. They prefer to eat meat and lay eggs in a nest. Snakes, lizards, crocodiles and turtles are reptiles.

DO: PLAYING WITH SCIENCE

SCIENTIFIC DEMONSTRATION: REPTILE EGG

Materials

- ☑ Clear glass
- ☑ Vinegar (NOTE—White vinegar works best, but you can also use apple cider vinegar.)
- ☑ Egg
- ☑ Plastic wrap
- ☑ Rubber band

Procedure

1. Gently place the egg in the cup.
2. Slowly pour the vinegar over the egg until the egg is completely covered.
3. Cover the cup with the plastic wrap and use the rubber band to secure it in place.
4. Set the cup aside in a safe place where it can sit undisturbed for twenty-four hours.
5. After a full day, take off the plastic wrap and use a spoon to fish out the egg. Rinse it gently and then let the students touch and feel the egg. Be careful though as it can easily break!

Explanation

The vinegar dissolved the calcium found in the hard shell of a chicken egg. This made the shell of the egg soft and leathery, just like a reptile's egg.

Take It Further

Have the students use their imaginations to make a snake out of clay.

(OPTIONAL) STEAM PROJECTS

Multi-chapter Activities

- ✂ FOOD CHART – This week, add the camel to the herbivore side and the cobra to the carnivore side of your food chart. You can use the mini-animal pictures found in the appendix on p. 129 of this guide or print out your own.

- ✂ HABITAT PROJECT – Make a poster or diorama that depicts the desert. This week, add the camel and the cobra. You can use the mini-animal pictures found in the appendix on p. 129 of this guide and the habitat poster on p. 131 or print out your own.

Activities For This Chapter

- ✂ EGG CARTON CAMEL – Make an egg carton camel. Paint three egg cups from an egg carton brown. Once they are dry, attach two brown pipe cleaners each to two of the egg cups for the legs of the camel. Then, attach the two egg cups together to create the camel's body. Then, attach one end of another brown pipe cleaner to the body and the other end to the final egg cup. Glue googly eyes on either side of the egg cup for the camel's head.

CHAPTER 5: GRID SCHEDULE

Supplies Needed	
Demo	• Hot water, 1 Coffee cup, 1 8"x10" Pan, Instant thermometer
Projects	• Newspaper, Flour, Salt, Pipe cleaners, Cheesecloth

Chapter Summary

The chapter opens with Tracey finally looping her rope over the column, so that the whole group is able to exit through the ventilation shaft. They escape the tomb through an underground river that leads to a well, where they see a spiny-tailed lizard fighting with a sand cat. Princess Talibah shares more about the lizard before the group sets out across the desert. They travel to a nomad camp nearby only to find that the Kekeway have already raided the village. They have taken the children because only the kids can fit into the dens of the fennec fox. Blaine, Tracey, Talibah, and her friends work with Mesneh, the Tuareg nomad, to teach Itja and his men a lesson. The chapter ends with the Sassafras twins, along with their companions, defeating the Kekeway band and rescuing the children.

Weekly Schedule

	Day 1	Day 2	Day 3	Day 4
Read	☐ Read the section entitled "Leaping Lizards" of Chapter 5 in *SSA Volume 1: Zoology.*	☐ Read the section entitled "Ferreting Out Foxes" of Chapter 5 in *SSA Volume 1: Zoology.*	☐ (*Optional*) Read one or all of the assigned pages from the encyclopedia of your choice.	☐ (*Optional*) Read one of the additional library books.
Write	☐ Fill out the Animal Record Sheet on SL p. 21 for the spiny-tailed lizard. ☐ Go over the vocabulary words and enter them into the Zoology Glossary on SL p. 95.	☐ Fill out the Animal Record Sheet on SL p. 22 for the fennec fox. ☐ Add to the Habitat Information Sheet on SL p. 17 for the Desert.	☐ Write the information learned from the demonstration on SL p. 24. ☐ (*Optional*) Write a narration on the Zoology Notes Sheet on SL p. 24.	☐ (*Optional*) Complete the copywork or dictation assignment and add it to the Zoology Notes sheet on SL p. 24. ☐ (*Optional*) Take Zoology Quiz #2.
Do	☐ (*Optional*) Make a Papier Maché lizard.	☐ (*Optional*) Learn about Endangered Animals.	☐ Do the demonstration entitled "Ear Cooling."	☐ (*Optional*) Add the animals studied this week to the food chart and habitat posters.

CHAPTER 5: LIST SCHEDULE

CHAPTER SUMMARY

The chapter opens with Tracey finally looping her rope over the column, so that the whole group is able to exit through the ventilation shaft. They escape the tomb through an underground river that leads to a well, where they see a spiny-tailed lizard fighting with a sand cat. Princess Talibah shares more about the lizard before the group sets out across the desert. They travel to a nomad camp nearby only to find that the Kekeway have already raided the village. They have taken the children because only the kids can fit into the dens of the fennec fox. Blaine, Tracey, Talibah, and her friends work with Mesneh, the Tuareg nomad, to teach Itja and his men a lesson. The chapter ends with the Sassafras twins, along with their companions, defeating the Kekeway band and rescuing the children.

ESSENTIALS

Read

☐ Read the section entitled "Leaping Lizards" of Chapter 5 in *SSA Volume 1: Zoology*.

☐ Read the section entitled "Ferreting Out Foxes" of Chapter 5 in *SSA Volume 1: Zoology*.

Write

☐ Fill out the Animal Record Sheet on SL p. 21 for the spiny-tailed lizard.

☐ Go over the vocabulary words and enter them into the Zoology Glossary on SL p. 95.

☐ Fill out the Animal Record Sheet on SL p. 22 for the fennec fox.

☐ Add to the Habitat Information Sheet on SL p. 17 for the Desert.

☐ Write the information learned from the demonstration on SL p. 24.

Do

☐ Do the demonstration entitled "Ear Cooling."

(OPTIONAL) EXTRAS

Read

☐ Read one of the additional library books.

☐ Read one or all of the assigned pages from the encyclopedia of your choice.

Write

☐ Write a narration on the Zoology Notes Sheet on SL p. 24.

☐ Complete the copywork or dictation assignment and add it to the Zoology Notes sheet on SL p. 24.

☐ Take Zoology Quiz #2.

Do

☐ Make a Papier Maché Lizard.

☐ Learn about Endangered Animals.

☐ Add the animals studied this week to the food chart and habitat posters.

Supplies Needed	
Demo	• Hot water, 1 Coffee cup, 1 8"x10" Pan, Instant thermometer
Projects	• Newspaper, Flour, Salt, Pipe cleaners, Cheesecloth

Chapter 5: Escaping the Tomb

Read: Gathering Information

Living Book Reading Assignment

📖 Chapter 5 of *The Sassafras Science Adventures Volume 1: Zoology*

(Optional) Encyclopedia Readings

🔖 *Kingfisher First Encyclopedia of Animals* p. 108 (Lizard), p. 33 (Fox)

🔖 *DK Encyclopedia of Animals* pp. 234-235 (Lizard), pp. 179-180 (Fox), pp. 68-69 (Deserts)

(Optional) Additional Library Books

📖 *Fun Facts About Lizards!* (I Like Reptiles and Amphibians!) by Carmen Bredeson

📖 *All About Lizards* by Jim Arnosky

📖 *Fox* by Kate Banks

📖 *Foxes* (Animal Predators) by Sandra Markle

📖 *Desert Animals* (Animals in Their Habitats) by Francine Galko

📖 *Life in the Desert* (Pebble Plus: Habitats Around the World) by Alison Auch

📖 *Baby Animals in Desert Habitats* (Habitats of Baby Animals) by Bobbie Kalman

Write: Keeping a Notebook

SCIDAT Logbook Sheets

This week, you can have the students fill out a logbook page for the spiny-tailed lizard and the fennec fox, as well as continue to fill out a Habitat Information Sheet and the Around the World Sheet for the Egyptian Desert. The students could use the following information:

Habitat Information Sheet - Desert

HABITAT: Desert

LOCATION AND LOCAL EXPERT

AVERAGE RAINFALL: There is very little rainfall.

AVERAGE TEMPERATURE: The desert has hot days and cold nights.

MAIN CHARACTERISTICS: There are rocky out-crops which provide shelter for plants and animals.

ANIMALS FOUND THERE: Add the spiny-tailed lizard and fennec fox (The students could also add scorpion and sand cat).

Around the World - Desert Habitat

MAP: Have the students put a star in the middle of the Egyptian Desert and label it "Egypt". Then, have them color the regions of the world that are covered with the desert habitat

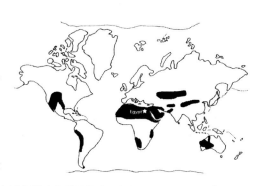

yellow. Use the map pictured as a guide.

CONTINENTS FOUND: North America, South America, Africa, Asia, Australia or Oceania

Animal Record Sheet - Lizard

ANIMAL NAME: Spiny-tailed Lizard

CLASSIFICATION: Reptile

FOOD: Omnivore

LOCATION FOUND: Egyptian Desert

INFORMATION LEARNED

- They are swift and agile animals.
- They live in rocky hill areas of the desert.
- Their scales are smooth, which helps them to burrow into their shelters.
- They are cold-blooded, so they need to spend time in the sun to keep warm.
- Their skin will darken in the sun to help their bodies absorb heat better.
- Some lizards will shed their tail when attacked and it will grow back over several months, but not the spiny-tailed lizard.
- It uses its spiked tail to defend itself against predators.
- The spiny-tailed lizard eats mostly plants, but will eat the occasional insect.
- A female can lay five to forty eggs at one time.

Animal Record Sheet - Fox

ANIMAL NAME: Fennec Fox

CLASSIFICATION: Mammal

FOOD: Carnivore

LOCATION FOUND: Egyptian Desert

INFORMATION LEARNED

- They are hunted for their furs.
- They are the smallest of the foxes, between one-and-a-half to four pounds, but they have the largest ears.
- Their large ears give them very good hearing, but also help with keeping cool.
- They have short legs, bushy tails, and fur-covered paws.
- They have sandy colored fur to blend in with the desert and reflect the sun's heat during the day. Their fur also keeps them warm at night.
- They live in burrows called dens.
- They are skillful hunters that come out at night and rest during the day.
- Their diet includes rodents, insects, birds, and eggs.
- They live in pairs or families that control a certain territory.
- The females are called vixens, males are dogs, and the young are cubs.
- Females give birth to live young and feed their young milk, which makes them mammals.

VOCABULARY

Have your older students look up the following term in the glossary in the appendix on pp. 145-146 or in a science encyclopedia. Have them copy the definition onto a blank index card or into their SCIDAT logbook.

☇ DESERT – A habitat that receives very little rain; it is hot during the day and cold at night.

(OPTIONAL) COPYWORK

Copywork Sentence

The desert has hot days and cool nights.

Dictation Selection

The desert has extremely hot days and chilly nights. It receives very little rain each year and is characterized by little to no life due to the lack of water. The plants that do survive are typically a variety of cacti.

(OPTIONAL) QUIZ

This week, you can give the students a quiz based on what they learned in chapters 4 and 5. You can find the quiz in the appendix on p. 157.

Quiz #2 Answers
1. B
2. B
3. A
4. C
5. B

DO: PLAYING WITH SCIENCE

SCIENTIFIC DEMONSTRATION: EAR COOLING

Materials
- ☑ Hot water
- ☑ 1 Coffee cup
- ☑ 1 8"x10" Pan
- ☑ Instant thermometer

Procedure
1. Fill the coffee cup and the pan with hot water. (CAUTION—Make sure that the water is not too hot to be touched.)
2. Then, take the temperature of the water in the coffee cup and the pan. Record the measurements as the initial temperatures.
3. After ten minutes, measure and record the temperature of the coffee cup and pan once more.
4. Repeat step 3, two more times for a total of thirty minutes.
5. Compare the temperatures you recorded to see if the water in the coffee cup cooled down at the same rate as the water in the pan.

Explanation

The students should see that the temperature of the water in the pan cooled quicker than the water in the coffee cup. The coffee cup insulates the water and prevents it from cooling down as quickly. The pan allows the water to spread out, giving it more contact with the air and thus more opportunity to cool off. The fennec fox has ears that are large, giving space for their blood to spread out near the skin. This proximity to the skin allows the blood plenty of room to cool, which helps the furry mammal to keep from overheating in the desert.

Take It Further

Have the students repeat the process with ice cold water. Does the water warm up in the same way that the hot water cooled down?

(OPTIONAL) STEAM PROJECTS

Multi-chapter Activities

✂ FOOD CHART — This week, add the spiny-tailed lizard to the omnivore side and the fennec fox to the carnivore side of your food chart. You can use the mini-animal pictures found in the appendix on p. 129 of this guide or print out your own.

✂ HABITAT PROJECT — This week, add the spiny-tailed lizard and fennec fox to your desert habitat. You can use the mini-animal pictures found in the appendix on p. 129 of this guide or print out your own.

Activities For This Chapter

✂ MAKE YOUR OWN LIZARD — Make a papier maché lizard using the directions from the video series found at the following website:

👆 http://www.ehow.com/videos-on_8268_children_s-crafts_-paper-mache-lizards.html

✂ ENDANGERED ANIMALS — At one point, the fennec fox was an endangered animal due to the fact that it was hunted for its fur. Explain to the students what being endangered means by saying:

> The fennec fox was once hunted for its fur so much that it was considered an endangered animal, which means that there were not very many of them left living in the wild. It also means that if we had not been careful, the fennec fox would have died out and we would have never been able to see them. Thanks to stricter hunting regulations and conservation efforts, the fennec fox has made a comeback, but many other species are still considered endangered.

If you know of an endangered species in your area, share that with the students. If not, you can visit the following website to find any endangered species that might live in your area.

👆 http://www.earthsendangered.com/list.asp

Try to learn a few facts about the animals and the current conservation efforts in your area.

CHAPTER 6: GRID SCHEDULE

Supplies Needed	
Demo	• 1 Pint of cream, 1 Large glass jar with lid, ½ Cup of water
Projects	• Air dry clay, Ingredients for cookies (white sugar, butter-flavored shortening, honey, eggs, vanilla extract, baking soda, all-purpose flour, and cinnamon)

Chapter Summary

The chapter opens with Blaine and Tracey zipping off to Canada. They run into Jethro (aka Jet) Mecklen and Edbert Snarfuffel, who are looking for a job on a farm. As the four walk down the road, they see that Smitty Farms is hiring, so they go talk to the boss. Jet impresses Farmer Smith with his knowledge of cows, so the farmer gives the group a trial day where they will have to work alongside his sons, Tank and Billy. Ed, the farmhand, takes them all out to the barn for their first job – milking the cows. Tank and Billy decide to pull a prank by knocking down a beehive from the rafters and locking the barn door. They succeed in disturbing the hive, but they also lock themselves into the barn with the others. The chapter ends with Jet sharing about the bees and his plan to save them all with an axe.

Weekly Schedule

	Day 1	Day 2	Day 3	Day 4
Read	☐ Read the section entitled "Compelling Cows" of Chapter 6 in *SSA Volume 1: Zoology.*	☐ Read the section entitled "Buzzing Bees!!" of Chapter 6 in *SSA Volume 1: Zoology.*	☐ (*Optional*) Read one or all of the assigned pages from the encyclopedia of your choice.	☐ (*Optional*) Read one of the additional library books.
Write	☐ Fill out the Animal Record Sheet on SL p. 29 for the cows. ☐ Go over the vocabulary words and enter them into the Zoology Glossary on SL p. 95.	☐ Fill out the Animal Record Sheet on SL p. 30 for the bee. ☐ Add to the Habitat Information Sheet on SL p. 27 for Farms.	☐ Write the information learned from the demonstration on SL p. 33. ☐ (*Optional*) Write a narration on the Zoology Notes Sheet on SL p. 33.	☐ (*Optional*) Complete the copywork or dictation assignment and add it to the Zoology Notes sheet on SL p. 33.
Do	☐ (*Optional*) Make a Cow Sculpture.	☐ (*Optional*) Make Bee Cookies.	☐ Do the demonstration entitled "Making Butter."	☐ (*Optional*) Add the animals studied this week to the food chart and habitat posters.

CHAPTER 6: LIST SCHEDULE

CHAPTER SUMMARY

The chapter opens with Blaine and Tracey zipping off to Canada. They run into Jethro (aka Jet) Mecklen and Edbert Snarfuffel, who are looking for a job on a farm. As the four walk down the road, they see that Smitty Farms is hiring, so they go talk to the boss. Jet impresses Farmer Smith with his knowledge of cows, so the farmer gives the group a trial day where they will have to work alongside his sons, Tank and Billy. Ed, the farmhand, takes them all out to the barn for their first job – milking the cows. Tank and Billy decide to pull a prank by knocking down a beehive from the rafters and locking the barn door. They succeed in disturbing the hive, but they also lock themselves into the barn with the others. The chapter ends with Jet sharing about the bees and his plan to save them all with an axe.

ESSENTIALS

Read

- ☐ Read the section entitled "Compelling Cows" of Chapter 6 in *SSA Volume 1: Zoology.*
- ☐ Read the section entitled "Buzzing Bees!!" of Chapter 6 in *SSA Volume 1: Zoology.*

Write

- ☐ Fill out the Animal Record Sheet on SL p. 29 for the cows.
- ☐ Go over the vocabulary words and enter them into the Zoology Glossary on SL p. 95.
- ☐ Fill out the Animal Record Sheet on SL p. 30 for the bee.
- ☐ Add to the Habitat Information Sheet on SL p. 27 for Farms.
- ☐ Write the information learned from the demonstration on SL p. 33.

Do

- ☐ Do the demonstration entitled "Making Butter."

(OPTIONAL) EXTRAS

Read

- ☐ Read one of the additional library books.
- ☐ Read one or all of the assigned pages from the encyclopedia of your choice.

Write

- ☐ Write a narration on the Zoology Notes Sheet on SL p. 33.
- ☐ Complete the copywork or dictation assignment and add it to the Zoology Notes sheet on SL p. 33.

Do

- ☐ Make a Cow Sculpture.
- ☐ Make Bee Cookies.
- ☐ Add the animals studied this week to the food chart and habitat posters.

Supplies Needed	
Demo	• 1 Pint of cream, 1 Large glass jar with lid, ½ Cup of water
Projects	• Air dry clay, Ingredients for cookies (white sugar, butter-flavored shortening, honey, eggs, vanilla extract, baking soda, all-purpose flour, and cinnamon)

CHAPTER 6: ON TO CANADA

READ: GATHERING INFORMATION

LIVING BOOK READING ASSIGNMENT

📖 Chapter 6 of *The Sassafras Science Adventures Volume 1: Zoology*

(OPTIONAL) ENCYCLOPEDIA READINGS

🔍 *Kingfisher First Encyclopedia of Animals* p. 77 (Cows), p. 137 (Bees)

🔍 *DK Encyclopedia of Animals* pp. 131-133 (Cows), pp. 111-112 (Bees)

(OPTIONAL) ADDITIONAL LIBRARY BOOKS

📖 *Cows and Their Calves* (Pebble Plus: Animal Offspring) by Margaret Hall

📖 *Raising Cows on the Koebels' Farm* (Our Neighborhood) by Alice K. Flanagan

📖 *Milk: From Cow to Carton* (Let's-Read-and-Find... Book) by Aliki

📖 *Are You a Bee?* (Backyard Books) by Judy Allen

📖 *The Life and Times of the Honeybee* by Charles Micucci

📖 *DK Readers: Busy, Buzzy Bee* (Level 1: Beginning to Read) by Karen Wallace

📖 *What if There Were No Bees?* by Suzanne Buckingham Slade and Carol Schwartz

WRITE: KEEPING A NOTEBOOK

SCIDAT LOGBOOK SHEETS

This week, you can have the students fill out a logbook page for the cow and the bee, as well as begin to fill out a Habitat Information Sheet for the Canadian Farm. The students could use the following information:

Habitat Information Sheet - Domestic Farm

HABITAT: Canadian Farm

LOCATION AND LOCAL EXPERT: Quebec, Canada and Jethro (or Jet) Mecklen

AVERAGE RAINFALL

AVERAGE TEMPERATURE: Farms in Canada have cold winters and warm summers with cool mornings.

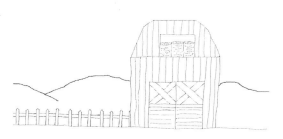

MAIN CHARACTERISTICS: They have rolling hills, pasture land, and mountains in the distance.

ANIMALS FOUND THERE: Add the cow and bee (The students could also add pig, horse, sheep, and turkey).

Animal Record Sheet - Cow

ANIMAL NAME: Cow

CLASSIFICATION: Mammal

FOOD: Herbivore

LOCATION FOUND: Canadian Farm

INFORMATION LEARNED

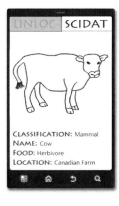

- There are over two hundred fifty breeds of cattle that are used all over the world for meat (beef) and milk.
- Cows are part of the bovine family.
- They are even toed, hoofed mammals that prefer to live in herds.
- They are herbivores and ruminants, which means they digest the grass they eat in stages within their four chambered stomachs.
- They have stocky bodies, wide heads, and strong legs; most have sharp sense of smell and vision.
- When cows are threatened, they can move with some speed.
- In some parts of the world, they are used to plow and pull carts.
- Holstein dairy cows can produce up to twenty-eight liters of milk per day.
- Only female cows produce milk, which was originally meant to feed their babies.
- They are milked twice a day by vacuum pumps that someone attaches to the udders, which takes about four minutes to finish.

Animal Record Sheet - Bee

ANIMAL NAME: Bee

CLASSIFICATION: Insect

FOOD: Herbivore

LOCATION FOUND: Canadian Farm

INFORMATION LEARNED

- Bees have black and yellow stripes, related to wasps.
- Only worker bees have stinging tails which they use to defend themselves.
- They usually die after they sting.
- Honeybees are ruled by a queen.
- She lays an egg in a wax room called a cell.
- The eggs grow and larvae hatch.
- The worker bees feed the larvae, it grows developing into a pupa, and after time, an adult bee emerges from the cell.
- They collect sweet juice from flowers called nectar and use it to make honey, which they store in the cells to feed the young.
- They are highly socialized insects and have sophisticated methods of communication, using dance to point the way to nectar rich flowers.
- Bees and wasps are the world's pollinators.
- *NOTE—The behavior of the bees in this chapter is not what we typically see in the wild. Typically, bees will build hives in the hollows of trees or in the walls, in other words in quiet spaces where they won't be disturbed. That said, it is possible for bees to build free-hanging hives, especially in abandoned buildings. Check out this article to learn more:
- https://www.honeybeesuite.com/a-free-hanging-honey-bee-nest-in-minnesota/

Vocabulary

Have your older students look up the following term in the glossary in the appendix on pp. 145-146 or in a science encyclopedia. Have them copy the definition onto a blank index card or into their SCIDAT logbook.

✍ **Domesticated Animal** – An animal that has been under human control for many generations.

(Optional) Copywork

Copywork Sentence

Cows, chickens, and pigs are domesticated animals.

Dictation Selection

Domestic farms can be found on every continent. A farm might raise crops, such as corn, or domesticated animals, such as cows. Domesticated animals, like chickens or pigs, have been under human control for many years.

Do: Playing with Science

Scientific Demonstration: Making Butter

Materials
- ☑ 1 Pint of cream
- ☑ 1 Large glass jar with lid
- ☑ ½ Cup of water

Procedure
1. Take the cream out the night before and let it sit on the counter for around 12 hours.
2. Pour the cream into the glass jar and use the lid to tightly seal the jar.
3. Hold the jar at the neck and shake vigorously with a downward thrusting motion for at least 10 minutes.
4. Have the students observe what happens. (NOTE—You should see a solid and a liquid in the jar.) Pour off the liquid buttermilk and then add ½ cup of water.
5. Shake the jar three to five more times and then pour off the liquid once more. The solid that is left is butter that you can use!

Explanation

Cows produce milk, which we use for a variety of products. If you allow the milk to sit a bit after you milk the cow, it will separate into skim milk and cream. The cream, which is full of butter fat, can be shaken, causing the fat molecules to break. The butter fat is then able to stick together to form butter, which we can eat.

Take It Further

Take the students on a field trip to visit a creamery or a dairy farm to see what other products can be make with milk and cream.

(Optional) Steam Projects

Multi-chapter Activities

✂ FOOD CHART — This week, add the cow and the bee to the herbivore side of your food chart. You can use the mini-animal pictures found in the appendix on p. 129 of this guide or print out your own.

✂ HABITAT PROJECT — Make a poster or diorama that depicts the domestic farm. This week, add the cow and bee. You can use the mini-animal pictures found in the appendix on p. 129 of this guide and the habitat poster on p. 132 or print out your own.

Activities For This Chapter

✂ COW SCULPTURE — Have the students sculpt a cow out of Air-dry clay. As the clay dries, look up the various breeds of Dairy Cattle in North America. You can check out a book from the local library or use the following website:

🖰 http://www.ansi.okstate.edu/breeds/cattle/

Have them choose a breed and read more about that breed. Once they finish, have them tell you (or write) several facts about the type of cow they chose. Finally, once their cow is dry, have them paint the cow the same color as their breed of choice.

✂ BEES COOKIES — Use the following recipe to make Honey Cookies:

- ☑ ½ Cup white sugar
- ☑ ¾ Cup butter flavored shortening
- ☑ 1 Cup honey
- ☑ 2 Eggs
- ☑ 1 Teaspoon vanilla extract
- ☑ 1 Teaspoon baking soda
- ☑ 3 ½ Cups all-purpose flour
- ☑ ½ Teaspoon ground cinnamon

Mix together the sugar, shortening, and honey in a saucepan over medium heat until the shortening melts. Set aside to cool. Mix together eggs, vanilla, baking soda, and cinnamon in another bowl. Add the egg mixture to the cooled honey mixture and then add the flour slowly. Drop a spoonful at a time onto a greased cookie sheet and bake at 350°F for 12 to 15 minutes. Let it cool before eating.

CHAPTER 7: GRID SCHEDULE

Supplies Needed	
Demo	• 1 Apple, 1 Glass jar
Projects	• Black pipe cleaners, Googly eyes, Styrofoam balls (one large, one small)

Chapter Summary

The chapter opens with Uncle Cecil recalling the twins' journey so far. Then the story moves to Blaine, Tracey, Jet, Edbert, Tank, and Billy escaping from the barn only to run into Ed, the farmhand. He takes them on to their next chore, which is to clean out the chicken coops. As they begin their work, Jet shares all he knows about chickens with the group. Tank and Billy decide that they are going to let Jet, Edbert, Blaine, and Tracey do all the work for the group while they take a quick nap. They wake up just before Ed returns with their lunch and to take them to their next job, which is to tear down an old barn infested with insects and spiders. Tank and Billy suggest a game of "Dare or Scare", which ends up with both of them being injured. Even so, the group is able to disassemble the barn before Ed Lumbia returns. He takes them all back to Farmer Smith, where the truth about what went on during the day comes out; Tank and Billy are reprimanded; Jet and Edbert are awarded jobs. Right before they leave, the twins find out that the Man With No Eyebrows has followed them to Canada.

Weekly Schedule

	Day 1	**Day 2**	**Day 3**	**Day 4**
Read	☐ Read the section entitled "Chicken Coop Cleaning" of Chapter 7 in *SSA Volume 1: Zoology*.	☐ Read the section entitled "Swinging with Spiders" of Chapter 7 in *SSA Volume 1: Zoology*.	☐ (*Optional*) Read one or all of the assigned pages from the encyclopedia of your choice.	☐ (*Optional*) Read one of the additional library books.
Write	☐ Fill out the Animal Record Sheet on SL p. 31 for the chicken. ☐ Go over the vocabulary words and enter them into the Zoology Glossary on SL p. 95.	☐ Fill out the Animal Record Sheet on SL p. 32 for the spider. ☐ Add to the Habitat Information Sheet on SL p. 27 for the Farms.	☐ Write the information learned from the demonstration on SL p. 34. ☐ (*Optional*) Write a narration on the Zoology Notes Sheet on SL p. 34.	☐ (*Optional*) Complete the copywork or dictation assignment and add it to the Zoology Notes sheet on SL p. 34. ☐ (*Optional*) Take Zoology Quiz #3.
Do	☐ (*Optional*) Take a Field Trip to a Local Farm.	☐ (*Optional*) Make a Spider.	☐ Do the demonstration entitled "Decomposing Insects."	☐ (*Optional*) Add the animals studied this week to the food chart and habitat posters.

CHAPTER 7: LIST SCHEDULE

CHAPTER SUMMARY

The chapter opens with Uncle Cecil recalling the twins' journey so far. Then the story moves to Blaine, Tracey, Jet, Edbert, Tank, and Billy escaping from the barn only to run into Ed, the farmhand. He takes them on to their next chore, which is to clean out the chicken coops. As they begin their work, Jet shares all he knows about chickens with the group. Tank and Billy decide that they are going to let Jet, Edbert, Blaine, and Tracey do all the work for the group while they take a quick nap. They wake up just before Ed returns with their lunch and to take them to their next job, which is to tear down an old barn infested with insects and spiders. Tank and Billy suggest a game of "Dare or Scare", which ends up with both of them being injured. Even so, the group is able to disassemble the barn before Ed Lumbia returns. He takes them all back to Farmer Smith, where the truth about what went on during the day comes out; Tank and Billy are reprimanded; Jet and Edbert are awarded jobs. Right before they leave, the twins find out that the Man With No Eyebrows has followed them to Canada.

ESSENTIALS

Read

☐ Read the section entitled "Chicken Coop Cleaning" of Chapter 7 in *SSA Volume 1: Zoology.*.

☐ Read the section entitled "Swinging with Spiders" of Chapter 7 in *SSA Volume 1: Zoology.*

Write

☐ Fill out the Animal Record Sheet on SL p. 31 for the chicken.

☐ Go over the vocabulary words and enter them into the Zoology Glossary on SL p. 95.

☐ Fill out the Animal Record Sheet on SL p. 32 for the spider.

☐ Add to the Habitat Information Sheet on SL p. 27 for the Farms.

☐ Write the information learned from the demonstration on SL p. 34.

Do

☐ Do the demonstration entitled "Decomposing Insects."

(OPTIONAL) EXTRAS

Read

☐ Read one of the additional library books.

☐ Read one or all of the assigned pages from the encyclopedia of your choice.

Write

☐ Write a narration on the Zoology Notes Sheet on SL p. 34.

☐ Complete the copywork or dictation assignment and add it to the Zoology Notes sheet on SL p. 34.

☐ Take Zoology Quiz #3.

Do

☐ Take a Field Trip to a Local Farm

☐ Make a Spider.

☐ Add the animals studied this week to the food chart and habitat posters.

Supplies Needed	
Demo	• 1 Apple, 1 Glass jar
Projects	• Black pipe cleaners, Googly eyes, Styrofoam balls (one large, one small)

CHAPTER 7: WORKING ON THE FARM

READ: GATHERING INFORMATION

LIVING BOOK READING ASSIGNMENT

📖 Chapter 7 of *The Sassafras Science Adventures Volume 1: Zoology*

(OPTIONAL) ENCYCLOPEDIA READINGS

🔦 *Kingfisher First Encyclopedia of Animals* p. 94 (Chicken), p. 133 (Spider), p. 134 (Insect)

🔦 *DK Encyclopedia of Animals* p. 138 (Chicken), pp. 327-328 (Spider), pp. 212-214 (Insect)

(OPTIONAL) ADDITIONAL LIBRARY BOOKS

📖 *From Egg to Chicken* (How Living Things Grow) by Anita Ganeri

📖 *Chickens Aren't the Only Ones* (World of Nature Series) by Ruth Heller

📖 *Chickens* (Animals That Live on the Farm) by JoAnn Early Macken

📖 *Time For Kids: Spiders!* by Editors of TIME For Kids

📖 *Spinning Spiders* (Let's-Read-and-Find... Science 2) by Melvin Berger

📖 *The Very Busy Spider* by Eric Carle

WRITE: KEEPING A NOTEBOOK

SCIDAT LOGBOOK SHEETS

This week, you can have the students fill out a logbook page for the chicken and the spider, as well as continue to fill out a Habitat Information Sheet and the Around the World Sheet for the Canadian Farm. The students could use the following information:

Habitat Information Sheet - Domestic Farm

HABITAT: Canadian Farm

LOCATION AND LOCAL EXPERT

AVERAGE RAINFALL

AVERAGE TEMPERATURE

MAIN CHARACTERISTICS

ANIMALS FOUND THERE: Add the chicken and spider.

Around the World - Continents

MAP: Have the students put a star in the middle of Quebec, Canada and label it "Quebec". Then have them label each of the seven continents and color all but Antarctica red, to show that domestic farms are found all over the world. Use the map pictured as a guide.

CONTINENTS FOUND: North America, South America, Africa, Asia, Europe, Australia or Oceania

Animal Record Sheet - Chicken

ANIMAL NAME: Chicken

CLASSIFICATION: Bird

FOOD: Omnivore

LOCATION FOUND: Canadian Farm

INFORMATION LEARNED

- Chickens are birds that can't fly very well.
- They prefer to run or walk.
- They are kept as farm animals all over the world.
- They are related to wild birds that were tamed by humans over four thousand years ago.
- Chickens develop a strict sense of seniority, called pecking order. The older chicken gets to eat first because they will peck the weaker members of the group to keep them in line.
- Male chickens are called roosters and they have large crests on their heads, a ruff of long feathers around their necks, and a long spike at the back of their leg called a spur.
- Male chickens crow at daybreak.
- Female chickens, called hens, are generally smaller and less colorful than roosters.
- Hens lay eggs.
- Females sit on their eggs until they hatch, getting up for less than an hour each day. They also turn the eggs periodically to ensure even incubation.
- Chicks will hatch after twenty-one days of incubation.
- Chicks have soft baby feathers that have different colors and patterns from their adult feathers; they will shed in a process called molting as they mature.

Animal Record Sheet - Spider

ANIMAL NAME: Spider

CLASSIFICATION: Invertebrate (or Arachnid)

FOOD: Carnivore

LOCATION FOUND: Canadian Farm

INFORMATION LEARNED

- Spiders are not insects.
- Spiders are part of the arachnid's class of invertebrates.
- They have a head and a rounded abdomen with eight legs attached.
- They are found all over the world.
- Spiders have poor eyesight.
- Most spiders have four pairs of simple eyes which work together to judge distances.
- They mainly feed on insects using their web to catch their prey.
- Their webs are made of silk which is very strong and stretches.
- The prey gets stuck in the web and the spider comes along, stuns it with venom from its fangs, and then wraps it in silk to eat it later.
- Silk is made from proteins produced in the abdomen, which are then drawn out of the body by its spinnerets.
- Spiders spin a new web every night and eat the old web to recycle the proteins.

Vocabulary

Have your older students look up the following terms in the glossary in the appendix on pp. 145-146 or in a science encyclopedia. Have them copy the definitions onto a blank index card or into their SCIDAT logbook.

- ✍ ARTHROPOD – An animal with a jointed body, such as an insect or spider.
- ✍ INSECT – An animal that has three jointed body parts (head, thorax, and abdomen) and six legs.

(Optional) Copywork

Copywork Sentence

Insects have six legs. Spiders have eight.

Dictation Selection

Arthropods are animals that have a jointed body. Insects and spiders are both arthropods, but they have a different number of legs. Insects have six legs, while spiders have eight.

(Optional) Quiz

This week, you can give the students a quiz based on what they learned in chapters 6 and 7. You can find the quiz in the appendix on p. 159.

Quiz #3 Answers
1. D
2. A
3. A
4. D
5. C

DO: Playing with Science

Scientific Demonstration: Insect Hunt

Materials
- ☑ A piece of fruit, such as an apple slice
- ☑ Honey or syrup
- ☑ Plate

Procedure
1. Have the students set the piece of fruit on half of the plate and squirt a bit of the honey or syrup on the other half.
2. Next, set the plate on the ground outside in a place that the students can easily see from the inside.
3. Let them observe the plate over the next hour or so, and record the insects they see.

Explanation

The students should see that the plate attracted several different insects. This demonstration was to help the student learn more about the insects in their area. As long as they have observed several insects, they have accomplished the purpose of this demonstration. You can also have the students use a field guide to identify and learn more about the insects they saw on the plate.

Take It Further

Have the students use a field guide to identify and learn more about the insects they saw on the plate.

(OPTIONAL) STEAM PROJECTS

Multi-chapter Activities

✂ FOOD CHART – This week, add the chicken to the omnivore side and spider to the carnivore side of your food chart. You can use the mini-animal pictures found in the appendix on p. 129 of this guide or print out your own.

✂ HABITAT PROJECT – This week, add the chicken and spider to your domestic farm habitat. You can use the mini-animal pictures found in the appendix on p. 129 of this guide or print out your own.

Activities For This Chapter

✂ FIELD TRIP – Take a field trip to a local farm that has chickens. Check out where the chickens live, what they eat and how the farm uses them.

✂ MAKE A SPIDER – Make a spider out of black pipe clears and Styrofoam balls. Begin by painting one small and one large Styrofoam ball black. Once they are dry, add googly eyes to the smaller Styrofoam ball for the head and attach it to the larger ball using a small piece of pipe cleaner. Then, cut 4 pipe cleaners in half and attach them to either side of the large Styrofoam ball for the eight spider legs.

CHAPTER 8: GRID SCHEDULE

Supplies Needed	
Demo	• 2-Liter Soda bottle with top, Gravel, Potting soil, Several small plants, Scissors, Tape, Water
Projects	• Toilet paper tube, Brown construction paper, Tissue paper (black, orange, yellow, white, brown, and green)

Chapter Summary

The chapter begins with Blaine and Tracey zipping off to Peru where they catch a boat for the "Out on a Limb" Guesthouse in the Amazon Rainforest. They are joined by their local expert, Alvaro, three hikers, a local man, and a dad with his two children. As they travel up river they spot a sloth which Alvaro shares more about. Once they arrive at the guesthouse, the twins meet the other guests and swing out to their room for the night. The next morning, Tracey has a strange encounter with a hummingbird in their room and again at breakfast. After they eat, the twins head out to a rainforest zip line along with Alvaro, Violetta, and Vancho. On the zip lines Blaine has a weird encounter with the same hummingbird. The group sees a some toucans, which Alvaro shares more about. The chapter ends with a group of chainsaw-carrying men approaching the very tree the zip-lining group is standing in.

Weekly Schedule

	Day 1	**Day 2**	**Day 3**	**Day 4**
Read	☐ Read the section entitled "Sloth Sighting" of Chapter 8 in *SSA Volume 1: Zoology*.	☐ Read the section entitled "Tree-hopping Toucans" of Chapter 8 in *SSA Volume 1: Zoology*.	☐ (*Optional*) Read one or all of the assigned pages from the encyclopedia of your choice.	☐ (*Optional*) Read one of the additional library books.
Write	☐ Fill out the Animal Record Sheet on SL p. 39 for the sloth. ☐ Go over the vocabulary words and enter them into the Zoology Glossary on SL p. 96.	☐ Fill out the Animal Record Sheet on SL p. 40 for the toucan. ☐ Add to the Habitat Information Sheet for the Rainforest on SL p. 37.	☐ Write the information learned from the demonstration on SL p. 43 ☐ (*Optional*) Write a narration on the Zoology Notes Sheet on SL p. 43.	☐ (*Optional*) Complete the copywork or dictation assignment and add it to the Zoology Notes sheet on SL p. 43.
Do	☐ (*Optional*) Make a Foot Sloth.	☐ (*Optional*) Make a Tissue Paper Toucan.	☐ Do the demonstration entitled "Rainforest in a Bottle."	☐ (*Optional*) Add the animals studied this week to the food chart and habitat posters.

CHAPTER 8: LIST SCHEDULE

CHAPTER SUMMARY

The chapter begins with Blaine and Tracey zipping off to Peru where they catch a boat for the "Out on a Limb" Guesthouse in the Amazon Rainforest. They are joined by their local expert, Alvaro, three hikers, a local man, and a dad with his two children. As they travel up river they spot a sloth which Alvaro shares more about. Once they arrive at the guesthouse, the twins meet the other guests and swing out to their room for the night. The next morning, Tracey has a strange encounter with a hummingbird in their room and again at breakfast. After they eat, the twins head out to a rainforest zip line along with Alvaro, Violetta, and Vancho. On the zip lines Blaine has a weird encounter with the same hummingbird. The group sees a some toucans, which Alvaro shares more about. The chapter ends with a group of chainsaw-carrying men approaching the very tree the zip-lining group is standing in.

ESSENTIALS

Read

☐ Read the section entitled "Sloth Sighting" of Chapter 8 in *SSA Volume 1: Zoology.*

☐ Read the section entitled "Tree-hopping Toucans" of Chapter 8 in *SSA Volume 1: Zoology.*

Write

☐ Fill out the Animal Record Sheet on SL p. 39 for the sloth.

☐ Go over the vocabulary words and enter them into the Zoology Glossary on SL p. 96.

☐ Fill out the Animal Record Sheet on SL p. 40 for the toucan.

☐ Add to the Habitat Information Sheet for the Rainforest on SL p. 37.

☐ Write the information learned from the demonstration on SL p. 43

Do

☐ Do the demonstration entitled "Rainforest in a Bottle."

(OPTIONAL) EXTRAS

Read

☐ Read one of the additional library books.

☐ Read one or all of the assigned pages from the encyclopedia of your choice.

Write

☐ Write a narration on the Zoology Notes Sheet on SL p. 43.

☐ Complete the copywork or dictation assignment and add it to the Zoology Notes sheet on SL p. 43.

Do

☐ Make a Foot Sloth.

☐ Make a Tissue Paper Toucan.

☐ Add the animals studied this week to the food chart and habitat posters.

Supplies Needed	
Demo	• 2-Liter Soda bottle with top, Gravel, Potting soil, Several small plants, Scissors, Tape, Water
Projects	• Toilet paper tube, Brown construction paper, Tissue paper (black, orange, yellow, white, brown, and green)

CHAPTER 8: ZIPPING THROUGH THE AMAZON RAINFOREST

READ: GATHERING INFORMATION

LIVING BOOK READING ASSIGNMENT

📖 Chapter 8 of *The Sassafras Science Adventures Volume 1: Zoology*

(OPTIONAL) ENCYCLOPEDIA READINGS

🔍 *Kingfisher First Encyclopedia of Animals* p. 89 (Toucan)
🔍 *DK Encyclopedia of Animals* pp. 321-322 (Sloth), p. 344 (Toucan), pp. 62-63 (Rainforest)

(OPTIONAL) ADDITIONAL LIBRARY BOOKS

📖 *"Slowly, Slowly, Slowly," said the Sloth* by Eric Carle
📖 *Baby Sloth* (Nature Babies) by Aubrey Lang and Wayne Lynch
📖 *Sloths* (Animals That Live in the Rain Forest) by Julie Guidone
📖 *Score One for the Sloths* by Helen Lester
📖 *Toucans and Other Birds* (Animals That Live in the Rain Forest) by Julie Guidone
📖 *Toucans* (Pebble Plus) by Mary R. Dunn
📖 *Toco Toucans: Bright Enough to Disappear* (Disappearing Acts) by Anastasia Suen
📖 *A Rainforest Habitat* (Introducing Habitats) by Molly Aloian and Bobbie Kalman
📖 *We're Roaming in the Rainforest: An Amazon Adventure* (Travel the World) by Laurie Krebs and Anne Wilson

WRITE: KEEPING A NOTEBOOK

SCIDAT LOGBOOK SHEETS

This week, you can have the students fill out a logbook page for the sloth and toucan, as well as begin to fill out a Habitat Information Sheet for the Amazon Rainforest. The students could use the following information:

Habitat Information Sheet - Rainforest

HABITAT: Amazon Rainforest

LOCATION AND LOCAL EXPERT: Peru and Alvaro Manihuari

AVERAGE RAINFALL: The rainforest has lots of rain.

AVERAGE TEMPERATURE: It is hot and humid.

MAIN CHARACTERISTICS: The rainforest is twice the size of the country of India, more than half the world's species of animals and plants live here, most animals and flowers and fruits are found up in the canopy, and the forest floor doesn't receive much sunlight.

ANIMALS FOUND THERE: Add the sloth and toucan. (The students could also add hummingbird and howler monkey.)

Animal Record Sheet - Sloth

ANIMAL NAME: Three-toed Sloth

CLASSIFICATION: Mammal

FOOD: Herbivore

LOCATION FOUND: Amazon Rainforest

INFORMATION LEARNED

- They spend their lives hanging upside down in trees by their hooked claws.
- They have three toes in the back and three toes in the front; each toe ends in a long curved claw.
- They sleep up to eighteen hours a day.
- They only climb to the forest floor once every one to two weeks to go to the bathroom.
- They are awake mostly at night.
- They feed on leaves and fruit.
- Sloths give birth to one young at a time.
- The mothers carry their baby sloth around on their stomachs for about five weeks.
- The baby sloth holds on by clinging to its mother's fur.
- They are covered in fur and are about the size of a small dog.
- Green algae grows on the coats of Amazonian sloths as a form of camouflage.
- They also have hundreds of beetles living in their coats that feed on the algae.
- Unlike other mammals, they don't regulate their body temperature internally.
- Sloths can rotate their heads through a two hundred seventy degree angle so that they can see all around them while hanging upside down.

Animal Record Sheet - Toucan

ANIMAL NAME: Toucan

CLASSIFICATION: Bird

FOOD: Omnivore

LOCATION FOUND: Amazon Rainforest

INFORMATION LEARNED

- Toucans are birds that live in the canopy of tropical rainforests in South America.
- They are some of the noisiest birds in the forest, as their call includes loud croaks, barks, and hoots.
- They are not excellent flyers, so they often walk or hop along branches, using their tails for balance.
- They are covered in black feathers, with a few white feathers around their eyes.
- Their large beaks are bright orange and yellow, but they are light because they are mostly hollow.
- They use their beaks to eat, to attract mates and to help in regulating their temperature.
- They use their beaks to pick fruit, which they toss back into their mouths and catch before they can eat.

- Passion fruit is their favorite food to eat.
- They also eat small birds or reptiles, which they pluck from cavities with their long bills.
- Their feet have two toes pointing forward and two toes pointing backward, making it easier for them to grip tree branches.
- They often nest in the hollows of decaying trees, laying clutches of two to five eggs in their nests.
- The male and female bird will both incubate the eggs until they have hatched.

VOCABULARY

Have your older students look up the following term in the glossary in the appendix on pp. 145-146 or in a science encyclopedia. Have them copy the definition onto a blank index card or into their SCIDAT logbook.

    ✐ RAINFOREST – A habitat with lots of plants, trees, and animals due to the heavy amount of rain it receives.

(OPTIONAL) COPYWORK

Copywork Sentence

The rainforest has a lot of different plants and animals.

Dictation Selection

"The Amazon rainforest is twice the size of the country of India, and more than half of the world's species of animals and plants live here. Most animals, flowers, and fruits are found up in the canopy, because the forest floor doesn't receive much sunlight," Alvaro shared.

DO: PLAYING WITH SCIENCE

SCIENTIFIC DEMONSTRATION: RAINFOREST IN A BOTTLE

Materials
- ☑ 2-Liter Soda bottle with top
- ☑ Gravel
- ☑ Potting soil
- ☑ Several small plants
- ☑ Scissors
- ☑ Tape
- ☑ Water

Procedure
1. Have an adult cut the soda bottle in half.
2. Pour in a layer of gravel about 1 inch thick on the bottom of one half. Cover the gravel with a layer of potting soil several inches deep.
3. Then plant your plants and water them well.
4. Tape the top half of the bottle back onto the bottom half and place the "Rainforest in a Bottle" on a sunny window sill.
5. Record what happens over the next several days.

Explanation

You should see that, after several hours, the bottle is coated with water droplets. Over several days, you will see that the soil remains moist and the bottle stays coated with water droplets. What is happening in the bottle is a small picture of the water cycle, which is repeated over and over in the rainforest.

Take It Further

Have the students build another terrarium habitat.

(OPTIONAL) STEAM PROJECTS

Multi-chapter Activities

✂ FOOD CHART – This week, add the sloth to the herbivore side and the toucan to the omnivore side of your food chart. You can use the mini-animal pictures found in the appendix on p. 129 of this guide or print out your own.

✂ HABITAT PROJECT – Make a poster or diorama that depicts the rainforest. This week, add the sloth and toucan. You can use the mini-animal pictures found in the appendix on p. 129 of this guide and the habitat poster on p. 133 or print out your own.

Activities For This Chapter

✂ FOOT SLOTH – Have the students make a sloth using paper, markers, tape and a toilet paper tube. Begin by tracing the students' foot onto the paper and cutting it out. This will be the sloth's body and head. Then cut out four rectangles for the sloth's arms and legs. Color all five pieces and then glue them together. Attach the sloth to the toilet paper tube so that it is hanging upside down. Add the sloth's claws and decorate the tube so that it looks like a branch of a tree.

✂ TISSUE PAPER TOUCAN – Have the students make a toucan using tissue paper. Begin by printing out the toucan template from the appendix on p. 138. Then glue balled up black, orange, yellow, white, brown, and green tissue paper to the toucan. Once dry, hang your toucan up to display.

CHAPTER 9: GRID SCHEDULE

	Supplies Needed	
Demo	• Life Cycle of a Butterfly Worksheet, Life Cycle of a Frog Worksheet	
Projects	• Paper (copy and construction)	

Chapter Summary

The chapter opens with Blaine, Tracey, Violetta, Vancho, and Alvaro having the tree cut down right out from under them. They all survive the fall and the rogue ProLog employees are scared off by poison arrows from a local tribe. Alvaro takes a moment to share some information about poison dart frogs before they notice a strange hummingbird flying near them. The twins realize that the hummingbird is a robot that is being operated by the Man With No Eyebrows. They chase after him, but he gets away just as the group is caught by a net set by the rogue ProLog employees. Alvaro shares about the blue-morpho butterflies that are hovering nearby to take the groups' minds off of their predicament. The natives once again rescue the group and the Perez children are reunited with their father. The chapter ends with an unknown man breaking into Uncle Cecil's basement laboratory.

Weekly Schedule

	Day 1	Day 2	Day 3	Day 4
Read	☐ Read the section entitled "Fearsome Frogs" of Chapter 9 in *SSA Volume 1: Zoology.*	☐ Read the section entitled "Beguiled by Butterflies" of Chapter 9 in *SSA Volume 1: Zoology.*	☐ (*Optional*) Read one or all of the assigned pages from the encyclopedia of your choice.	☐ (*Optional*) Read one of the additional library books.
Write	☐ Fill out the Animal Record Sheet on SL p. 41 for the frog. ☐ Go over the vocabulary words and enter them into the Zoology Glossary on SL p. 96.	☐ Fill out the Animal Record Sheet on SL p. 42 for the butterfly. ☐ Add to the Habitat Information Sheet on SL p. 37 for the Rainforest.	☐ Write the information learned from the demonstration on SL p. 44. ☐ (*Optional*) Write a narration on the Zoology Notes Sheet on SL p. 44.	☐ (*Optional*) Complete the copywork or dictation assignment and add it to the Zoology Notes sheet on SL p. 44. ☐ (*Optional*) Take Zoology Quiz #4.
Do	☐ (*Optional*) Have a Frog Race.	☐ (*Optional*) Create Butterfly Symmetry.	☐ Do the demonstration entitled "Examining Life Cycles."	☐ (*Optional*) Add the animals studied this week to the food chart and habitat posters.

CHAPTER 9: LIST SCHEDULE

CHAPTER SUMMARY

The chapter opens with Blaine, Tracey, Violetta, Vancho, and Alvaro having the tree cut down right out from under them. They all survive the fall and the rogue ProLog employees are scared off by poison arrows from a local tribe. Alvaro takes a moment to share some information about poison dart frogs before they notice a strange hummingbird flying near them. The twins realize that the hummingbird is a robot that is being operated by the Man With No Eyebrows. They chase after him, but he gets away just as the group is caught by a net set by the rogue ProLog employees. Alvaro shares about the blue-morpho butterflies that are hovering nearby to take the groups' minds off of their predicament. The natives once again rescue the group and the Perez children are reunited with their father. The chapter ends with an unknown man breaking into Uncle Cecil's basement laboratory.

ESSENTIALS

Read

☐ Read the section entitled "Fearsome Frogs" of Chapter 9 in *SSA Volume 1: Zoology*.

☐ Read the section entitled "Beguiled by Butterflies" of Chapter 9 in *SSA Volume 1: Zoology*.

Write

☐ Fill out the Animal Record Sheet on SL p. 41 for the frog.

☐ Go over the vocabulary words and enter them into the Zoology Glossary on SL p. 96.

☐ Fill out the Animal Record Sheet on SL p. 42 for the butterfly.

☐ Add to the Habitat Information Sheet on SL p. 37 for the Rainforest.

☐ Write the information learned from the demonstration on SL p. 44.

Do

☐ Do the demonstration entitled "Examining Life Cycles."

(OPTIONAL) EXTRAS

Read

☐ Read one of the additional library books.

☐ Read one or all of the assigned pages from the encyclopedia of your choice.

Write

☐ Write a narration on the Zoology Notes Sheet on SL p. 44.

☐ Complete the copywork or dictation assignment and add it to the Zoology Notes sheet on SL p. 44.

☐ Take Zoology Quiz #4

Do

☐ Have a Frog Race.

☐ Create Butterfly Symmetry.

☐ Add the animals studied this week to the food chart and habitat posters.

Supplies Needed	
Demo	• Life Cycle of a Butterfly Worksheet, Life Cycle of a Frog Worksheet
Projects	• Paper (copy and construction)

Chapter 9: Trouble in the Jungle

READ: Gathering Information

Living Book Reading Assignment

📖 Chapter 9 of *The Sassafras Science Adventures Volume 1: Zoology*

(Optional) Encyclopedia Readings

📍 *Kingfisher First Encyclopedia of Animals* p. 115 (Frog), p. 114 (Amphibian), p. 138 (Butterfly)

📍 *DK Encyclopedia of Animals* pp. 181-182 (Frog), pp. 92-94 (Amphibian), pp. 121-123 (Butterfly)

(Optional) Additional Library Books

📖 *From Tadpole to Frog* (Let's-Read-and-Find... Science 1) by Wendy Pfeffer

📖 *Frogs and Toads and Tadpoles, Too* (Rookie Read-About Science) by Allan Fowler

📖 *National Geographic Readers: Frogs!* by Elizabeth Carney

📖 *From Caterpillar to Butterfly* (Let's-Read-and-Find...) by Deborah Heiligman

📖 *National Geographic Readers: Great Migrations Butterflies* by Laura F. Marsh

📖 *Caterpillars and Butterflies* (Usborne Beginners) by Stephanie Turnbull

WRITE: Keeping a Notebook

SCIDAT Logbook Sheets

This week, you can have the students fill out a logbook page for the frog and the butterfly, as well as continue to fill out a Habitat Information Sheet and the Around the World Sheet for the Amazon Rainforest. The students could use the following information:

Habitat Information Sheet - Rainforest

HABITAT: Amazon Rainforest

LOCATION AND LOCAL EXPERT

AVERAGE RAINFALL

AVERAGE TEMPERATURE

MAIN CHARACTERISTICS

ANIMALS FOUND THERE: Add the frog and butterfly.

Around the World - Rainforest Habitat

MAP: Have the students put a star in the top of Peru, where the Amazon Rainforest is, and label it "Peru". Then, have them color the regions of the world that are covered with the rainforest habitat dark green. Use the map pictured as a guide.

CONTINENTS FOUND: North America, South America, Africa, Asia, Australia or Oceania

Animal Record Sheet - Frog

ANIMAL NAME: Poison Dart Frog

CLASSIFICATION: Amphibian

FOOD: Carnivore

LOCATION FOUND: Amazon Rainforest

INFORMATION LEARNED

- The students could share information about the frog life cycle (see worksheet in the appendix on p. 139 for answers).
- They are brightly colored orange, yellow, red, green, and blue.
- They eat fruit flies, ants, termites, small crickets, and tiny beetles, which they catch with their long sticky tongues and then swallow whole.
- They are amphibians with strong back legs that they use for jumping.
- They are also very good swimmers.
- The Amazon snake has developed a tolerance for the alkaloid poison secreted by the frogs, so it is able to eat the amphibians.
- They secrete an alkaloid toxin slime that covers their body.
- Natives use the most poisonous varieties of their slime on the tips of their arrows when hunting.
- Frogs call by moving air across a series of vocal cords in their inflatable throat pouch.

Animal Record Sheet - Butterfly

ANIMAL NAME: Blue-morpho Butterfly

CLASSIFICATION: Insect

FOOD: Herbivore

LOCATION FOUND: Amazon Rainforest

INFORMATION LEARNED

- Blue Morpho has bright blue wings that are edged with black, but the underside of their wings are dull brown with several eyespots as camouflage.
- Every part of their body is covered by thousands of tiny scales.
- They have two forewings and two hind wings, three body parts (head, thorax, and abdomen), and six legs, which makes them insects.
- They sometimes head to the canopy for mating or to sun themselves.
- They spend most of their time on the forest floor eating rotting fruits, tree sap, fungi, wet mud, and the juice from decomposing plants or animals.
- They feed entirely on liquids, which they suck up through their hollow tongues, called a proboscis.
- When not using their tongue they coil it up like a spring under their head.
- They have antennae, which they use to detect scents in the air of food or other members of their species.
- The students could share information about the butterfly life cycle (see worksheet in the appendix on p. 141 for answers).
- Blue Morpho caterpillars feed at night on plants that are a part of the pea family that can contain a toxin which the butterfly will secrete later, making it toxic to animals.

- Most butterflies are colorful and fly by day, most moths fly at night and are dull in color; most butterflies rest with their wings up, while moths rest with their wings flat.
- Butterflies generally have long slender antennae with clubbed ends while moths can have feathered antennae.

VOCABULARY

Have your older students look up the following term in the glossary in the appendix on pp. 145-146 or in a science encyclopedia. Have them copy the definition onto a blank index card or into their SCIDAT logbook.

✎ AMPHIBIAN – A cold-blooded, smooth-skinned vertebrate, such as a frog or salamander.

(OPTIONAL) COPYWORK

Copywork Sentence

Amphibians, like frogs, are cold-blooded.

Dictation Selection

The word amphibian comes from the Greek words, "amphi" and "bios", meaning double life. Amphibians are cold-blooded, smooth-skinned vertebrates, like frogs or salamanders. Their young typically begin life in water, but once they grow lungs, they live on land.

(OPTIONAL) QUIZ

This week, you can give the students a quiz based on what they learned in chapters 8 and 9. You can find the quiz in the appendix on p. 161.

Quiz #4 Answers
1. B
2. A
3. D
4. A
5. C

DO: PLAYING WITH SCIENCE

SCIENTIFIC DEMONSTRATION: EXAMINING LIFE CYCLES

Materials
- ☑ Life Cycle of a Butterfly Worksheet
- ☑ Life Cycle of a Frog Worksheet

Procedure
1. Explain to the students the life cycle of a butterfly and frog using the worksheets provided in the appendix on pp. 139-140.
2. If you have an older students, have them fill in the blank forms in the appendix on pp. 141-142 as you go along. If you have a younger students, let them color the completed worksheets as you explain them.

Take It Further

Have the students observe a life cycle by raising a tadpole into a frog or by watching a caterpillar grow into a butterfly.

(Optional) STEAM Projects

Multi-chapter Activities

- ✂ FOOD CHART – This week, add the frog to the carnivore side and the butterfly to the herbivore side of your food chart. You can use the mini-animal pictures found in the appendix on p. 129 of this guide or print out your own.

- ✂ HABITAT PROJECT – This week, add the frog and butterfly to your rainforest habitat. You can use the mini-animal pictures found in the appendix on p. 129 of this guide or print out your own.

Activities For This Chapter

- ✂ FROG RACE – Have the students race to the finish line while hopping like frogs.

- ✂ BUTTERFLY SYMMETRY – Have the students paint one half of a sheet of paper as they choose. While the paint is wet, fold sheet in half to get two equal sides. Cut out a butterfly shape and paste it onto a sheet of construction paper.

CHAPTER 10: GRID SCHEDULE

	Supplies Needed
Demo	• 2 Thermometers, Large felt rectangle, Tape, Plastic baggie (sealable), Warm water
Projects	• Ingredients for Hot-maybe eggs (mushrooms, oil, tomato, eggs, milk, and cream cheese), Fused beads, Empty can, Cotton balls, Felt (pink and blue)

Chapter Summary

The chapter opens with the mysterious Man With No Eyebrows corrupting the twins' SCIDAT data. As a result, Blaine zips to Australia without Tracey. He meets his local expert Willy Day in a local diner over a plate of "hot-maybe" eggs served by Ethel. The two head out for the Brown Mountain Forest where Willy is filming a documentary about the forest's inhabitants, including the infamous Feuding Brown Mountain Hermits. On the way into the forest, they see several koalas that Willy shares about in his documentary. As night falls, the forest comes alive with strange noises and a band of rabbits attacks. Willy shares all about rabbits including the fact that their presence is probably a sign of one of the infamous Feuding Brown Mountain Hermits. The chapter ends with Willy crying out to Ralphy Dingo.

Weekly Schedule

	Day 1	**Day 2**	**Day 3**	**Day 4**
Read	☐ Read the section entitled "Capturing Koalas" of Chapter 10 in *SSA Volume 1: Zoology.*	☐ Read the section entitled "Rampant Rabbits" of Chapter 10 in *SSA Volume 1: Zoology.*	☐ (*Optional*) Read one or all of the assigned pages from the encyclopedia of your choice.	☐ (*Optional*) Read one of the additional library books.
Write	☐ Fill out the Animal Record Sheet on SL p. 49 for the koala. ☐ Go over the vocabulary words and enter them into the Zoology Glossary on SL p. 96.	☐ Fill out the Animal Record Sheet on SL p. 50 for the rabbit. ☐ Add to the Habitat Information Sheet for the Eucalyptus Forest on SL p. 47.	☐ Write the information learned from the demonstration on SL p. 53. ☐ (*Optional*) Write a narration on the Zoology Notes Sheet on SL p. 53.	☐ (*Optional*) Complete the copywork or dictation assignment and add it to the Zoology Notes sheet on SL p. 53.
Do	☐ (*Optional*) Make Hot-Maybe Eggs. ☐ (*Optional*) Make a Koala.	☐ (*Optional*) Make a Rabbit.	☐ Do the demonstration entitled "Pouch Living."	☐ (*Optional*) Add the animals studied this week to the food chart and habitat posters.

CHAPTER 10: LIST SCHEDULE

CHAPTER SUMMARY

The chapter opens with the mysterious Man With No Eyebrows corrupting the twins' SCIDAT data. As a result, Blaine zips to Australia without Tracey. He meets his local expert Willy Day in a local diner over a plate of "hot-maybe" eggs served by Ethel. The two head out for the Brown Mountain Forest where Willy is filming a documentary about the forest's inhabitants, including the infamous Feuding Brown Mountain Hermits. On the way into the forest, they see several koalas that Willy shares about in his documentary. As night falls, the forest comes alive with strange noises and a band of rabbits attacks. Willy shares all about rabbits including the fact that their presence is probably a sign of one of the infamous Feuding Brown Mountain Hermits. The chapter ends with Willy crying out to Ralphy Dingo.

ESSENTIALS

Read

☐ Read the section entitled "Capturing Koalas" of Chapter 10 in *SSA Volume 1: Zoology.*

☐ Read the section entitled "Rampant Rabbits" of Chapter 10 in *SSA Volume 1: Zoology.*

Write

☐ Fill out the Animal Record Sheet on SL p. 49 for the koala.

☐ Go over the vocabulary words and enter them into the Zoology Glossary on SL p. 96.

☐ Fill out the Animal Record Sheet on SL p. 50 for the rabbit.

☐ Add to the Habitat Information Sheet for the Eucalyptus Forest on SL p. 47.

☐ Write the information learned from the demonstration on SL p. 53.

Do

☐ Do the demonstration entitled "Pouch Living."

(OPTIONAL) EXTRAS

Read

☐ Read one of the additional library books.

☐ Read one or all of the assigned pages from the encyclopedia of your choice.

Write

☐ Write a narration on the Zoology Notes Sheet on SL p. 53.

☐ Complete the copywork or dictation assignment and add it to the Zoology Notes sheet on SL p. 53.

Do

☐ Make Hot-Maybe Eggs.

☐ Make a Koala.

☐ Add the animals studied this week to the food chart and habitat posters.

Supplies Needed	
Demo	• 2 Thermometers, Large felt rectangle, Tape, Plastic baggie (sealable), Warm water
Projects	• Ingredients for Hot-maybe eggs (mushrooms, oil, tomato, eggs, milk, and cream cheese), Fused beads, Empty can, Cotton balls, Felt (pink and blue)

CHAPTER 10: DIVERGING TO AUSTRALIA

READ: GATHERING INFORMATION

LIVING BOOK READING ASSIGNMENT

📖 Chapter 10 of *The Sassafras Science Adventures Volume 1: Zoology*

(OPTIONAL) ENCYCLOPEDIA READINGS

🔖 *Kingfisher First Encyclopedia of Animals* p. 50 (Koala), p. 65 (Rabbit)

🔖 *DK Encyclopedia of Animals* p. 226 (Koala), pp. 288-289 (Rabbit)

(OPTIONAL) ADDITIONAL LIBRARY BOOKS

📖 *Koala* (Life Cycle of A...) by Bobbie Kalman

📖 *A Koala Is Not a Bear!* (Crabapples) by Hannelore Sotzek

📖 *Rabbits* (Blastoff! Readers: Backyard Wildlife) by Derek Zobel

📖 *Rabbits and Raindrops* by Jim Arnosky

📖 *The Little Rabbit* by Judy Dunn

📖 *The Tale of Peter Rabbit* by Beatrix Potter

WRITE: KEEPING A NOTEBOOK

SCIDAT LOGBOOK SHEETS

This week, you can have the students begin to fill out a Habitat Information Sheet for the Eucalyptus Forest and a logbook page for koala and rabbit. The students could use the following information:

Habitat Information Sheet - Forest

HABITAT: Brown Mountain Forest (or Eucalyptus Forest)

LOCATION AND LOCAL EXPERT: Victoria, Australia and Willy Day

AVERAGE RAINFALL

AVERAGE TEMPERATURE

MAIN CHARACTERISTICS: The Brown Mountain Forest is an old growth forest with lots of eucalyptus trees.

ANIMALS FOUND THERE: Add the koala and rabbit (The students could also add spot-tailed quell and long-footed potaroo).

Animal Record Sheet - Koala

ANIMAL NAME: Koala

CLASSIFICATION: Mammal

FOOD: Herbivore

LOCATION FOUND: Brown Mountain Forest

INFORMATION LEARNED

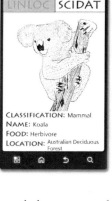

- Koalas live in the eucalyptus forests.
- Koalas look like bears, so they are called koala bears, but they are really marsupials.
- Marsupials are mammals, with the exception that their young are born in an immature state.
- Baby koalas complete their development in their mother's pouch.
- When the babies are born they are roughly the size of bees.
- They crawl up their mother's abdomen and into the pouch where they continue their development.
- When they are ready to leave the pouch they climb onto their mother's back until they are ready to go out on their own.
- They spend most of their lives up in the eucalyptus trees, eating the leaves and bark.
- They only come down to cross to another clump of trees.
- They have strong legs and sharp claws that they use for gripping tree trunks; they climb by bringing up their hind legs in a series of jumps.
- They have to rest for up to eighteen hours a day because their diet is not very nutritious.

Animal Record Sheet - Rabbit

ANIMAL NAME: Rabbit

CLASSIFICATION: Mammal

FOOD: Herbivore

LOCATION FOUND: Brown Mountain Forest

INFORMATION LEARNED

- They are mammals that feed mainly on grass and plants.
- Rabbits are an invasion species, meaning they are not natural to Australia.
- They were introduced by Europeans in 1788 and they are responsible for the loss of many plant species.
- Rabbits were originally only found in the Mediterranean, but humans have now introduced them all over the world.
- Rabbits dig burrows to provide protection from predators and shelter from the weather.
- They also give birth in their burrows.
- They can have up to nine babies in a litter and have seven liters per year.
- The babies are born hairless and blind, so the mother keeps them in an underground nest for around three weeks.
- They have large hind legs that they use for bounding away and for speed when running from predators.

VOCABULARY

Have your older students look up the following term in the glossary in the appendix on pp. 145-146 or in a science encyclopedia. Have them copy the definition onto a blank index card or into their SCIDAT logbook.

- ↻ MARSUPIAL – A group of mammals that give birth to immature young that complete their development in the mother's pouch.

(Optional) Copywork

Copywork Sentence

Kangaroos and koalas are marsupials.

Dictation Selection

Marsupials, like all other mammals, give birth to live young. However, marsupials give birth to immature young who finish their growth in their mother's pouch. Marsupials include kangaroos, opossums, wallaby, Tasmanian devils, and wombats.

DO: Playing with Science

Scientific Demonstration: Pouch Living

Materials
- ☑ 2 Thermometers
- ☑ Large felt rectangle
- ☑ Tape
- ☑ Plastic baggie (sealable)
- ☑ Warm water

Procedure
1. Set both the thermometers on the counter and wait 5 minutes. Then record the temperature of each.
2. Fold the felt rectangle in half and tape it on the sides to create a pouch. Then fill a plastic baggie with water warm to the touch and seal it.
3. Place the baggie flat on the counter and lay the felt pouch over it. Then place one of the thermometers into the felt pouch.
4. After 5 minutes, record the temperature of the two thermometers once more. Then wait 5 more minutes and record the two temperatures again.

Explanation

The students should see that the thermometer in the pouch increased in temperature. One of the main reasons that marsupial babies continue their growth inside their mother's pouch is because they cannot produce enough heat to keep themselves alive. This is because the majority of their energy is going towards their development. Also, they lack fur, so they must remain in the pouch to keep warm.

Take It Further

Have the students repeat the demonstration with ice-cold water in the plastic baggie to see if the results in the changes in temperature remain the same.

(Optional) STEAM Projects

Multi-chapter Activities

✂ FOOD CHART – This week, add the koala and the rabbit to the herbivore side of your food chart. You can use the mini-animal pictures found in the appendix on p. 129 of this guide or print out your own.

✂ HABITAT PROJECT ~ Make a poster or diorama that depicts the forest. This week, add the koala and rabbit. You can use the mini-animal pictures found in the appendix on p. 129 of this guide and the habitat poster on p. 138 or print out your own.

Activities For This Chapter

✂ HOT-MAYBE EGGS ~ Use the recipe below to make your own hot-maybe eggs.

- ☑ 4-6 Mushrooms, chopped
- ☑ 1 tsp Vegetable oil
- ☑ ½ Tomato, chopped
- ☑ A pinch each of salt, pepper and garlic powder
- ☑ 4 Eggs
- ☑ ¼ Cup of milk
- ☑ 2 Tbsp Cream cheese

Add the oil and mushrooms into a skillet and cook on medium heat until the mushrooms are almost done. Meanwhile, crack the eggs into a bowl, add the milk, and whisk until well combined. Once the mushrooms are almost done, add the tomatoes and seasonings. Cook for 1-2 minutes before adding the egg mixture. Using a spatula, move the eggs around until they are completely cooked and then remove from the heat. Add the cream cheese in several dollops, stir well and serve.

✂ MAKE A KOALA ~ Make a fused bead koala from the pattern found here:

🖱 http://www.activityvillage.co.uk/koala_fuse_bead_pattern.htm

NOTE—If the students know how to cross-stitch you could have them use the above pattern for a cross-stitch project.

✂ MAKE A RABBIT ~ Make a rabbit out of a can by painting the can white. Once it dries, cover it with glue or spray adhesive and glue cotton balls all over the can. Glue pink ears, blue eyes and a pink nose to the end of the can. Then, glue 4 white feet to the side of the can. Finally, draw a face using a permanent marker below the nose and add a few pipe cleaner whiskers.

CHAPTER 11: GRID SCHEDULE

Supplies Needed	
Demo	• Chopsticks, Tweezers, Pliers, Eye dropper, Sugar water or honey, Gummy worms, Unshelled peanuts, Seeds, Raisins, Plate
Projects	• Felt (white and black), Googly eyes, Paper (copy and construction)

Chapter Summary

The chapter opens with Tracey zipping off to China in hopes of finding Blaine. However, once she gets there, Uncle Cecil lets her know that she must complete this leg of the journey on her own. She meets her local expert, Tashi Yidro, and they travel up into the mountains on a bus. They stop at the Panda Reserve where Tashi shares about pandas just before a fire-cracker explodes nearby. The two girls are blamed for the explosion and are asked to leave the park. They continue on the bus towards Tashi's hometown, but on the way the bus has an accident caused by some more fireworks and the girls have to continue on foot. As they are walking they see an eagle flying up high and Tashi shares a little about them. The chapter ends with another sound like an explosion, only this time it's a landslide!

Weekly Schedule

	Day 1	Day 2	Day 3	Day 4
Read	☐ Read the section entitled "Panda Park" of Chapter 11 in *SSA Volume 1: Zoology*.	☐ Read the section entitled "Explosions and Eagles" of Chapter 11 in *SSA Volume 1: Zoology*.	☐ (*Optional*) Read one or all of the assigned pages from the encyclopedia of your choice.	☐ (*Optional*) Read one of the additional library books.
Write	☐ Fill out the Animal Record Sheet on SL p. 59 for the panda. ☐ Go over the vocabulary words and enter them into the Zoology Glossary on SL p. 97.	☐ Fill out the Animal Record Sheet on SL p. 60 for the eagle. ☐ Add to the Habitat Information Sheet for the Bamboo Forest on SL p. 57.	☐ Write the information learned from the demonstration on SL p. 63. ☐ (*Optional*) Write a narration on the Zoology Notes Sheet on SL p. 63.	☐ (*Optional*) Complete the copywork or dictation assignment and add it to the Zoology Notes sheet on SL p. 63.
Do	☐ (*Optional*) Make a Panda.	☐ (*Optional*) Make an Eagle.	☐ Do the demonstration entitled "Bird Beaks."	☐ (*Optional*) Add the animals studied this week to the food chart and habitat posters.

CHAPTER 11: LIST SCHEDULE

CHAPTER SUMMARY

The chapter opens with Tracey zipping off to China in hopes of finding Blaine. However, once she gets there, Uncle Cecil lets her know that she must complete this leg of the journey on her own. She meets her local expert, Tashi Yidro, and they travel up into the mountains on a bus. They stop at the Panda Reserve where Tashi shares about pandas just before a fire-cracker explodes nearby. The two girls are blamed for the explosion and are asked to leave the park. They continue on the bus towards Tashi's hometown, but on the way the bus has an accident caused by some more fireworks and the girls have to continue on foot. As they are walking they see an eagle flying up high and Tashi shares a little about them. The chapter ends with another sound like an explosion, only this time it's a landslide!

ESSENTIALS

Read

☐ Read the section entitled "Panda Park" of Chapter 11 in *SSA Volume 1: Zoology*.

☐ Read the section entitled "Explosions and Eagles" of Chapter 11 in *SSA Volume 1: Zoology*.

Write

☐ Fill out the Animal Record Sheet on SL p. 59 for the panda.

☐ Go over the vocabulary words and enter them into the Zoology Glossary on SL p. 97.

☐ Fill out the Animal Record Sheet on SL p. 60 for the eagle.

☐ Add to the Habitat Information Sheet for the Bamboo Forest on SL p. 57.

☐ Write the information learned from the demonstration on SL p. 63.

Do

☐ Do the demonstration entitled "Bird Beaks."

(OPTIONAL) EXTRAS

Read

☐ Read one of the additional library books.

☐ Read one or all of the assigned pages from the encyclopedia of your choice.

Write

☐ Write a narration on the Zoology Notes Sheet on SL p. 63.

☐ Complete the copywork or dictation assignment and add it to the Zoology Notes sheet on SL p. 63.

Do

☐ Make a Panda.

☐ Make an Eagle.

☐ Add the animals studied this week to the food chart and habitat posters.

Supplies Needed	
Demo	• Chopsticks, Tweezers, Pliers, Eye dropper, Sugar water or honey, Gummy worms, Unshelled peanuts, Seeds, Raisins, Plate
Projects	• Felt (white and black), Googly eyes, Paper (copy and construction)

Chapter 11: Separated in China

Read: Gathering Information

Living Book Reading Assignment

📖 Chapter 11 of *The Sassafras Science Adventures Volume 1: Zoology*

(Optional) Encyclopedia Readings

🔖 *Kingfisher First Encyclopedia of Animals* p. 41 (Panda), p. 85 (Eagle), p. 84 (Bird)

🔖 *DK Encyclopedia of Animals* pp. 270-272 (Panda), pp. 167-168 (Eagle), pp. 115-117 (Bird)

(Optional) Additional Library Books

📖 *Pi-Shu the Little Panda* by John Butler

📖 *Endangered Pandas* (Earth's Endangered Animals) by John Crossingham

📖 *Tracks of a Panda* by Nick Dowson

📖 *Eagles* (Animal Predators) by Sandra Markle

📖 *Bald Eagles* (Nature Watch (Lerner)) by Charlotte Wilcox

📖 *Challenger: America's Favorite Eagle* by Margot Theis Raven

Write: Keeping a Notebook

SCIDAT Logbook Sheets

This week, you can have the students begin to fill out a Habitat Information Sheet for the Bamboo Forest and a logbook page for the panda and the eagle. The students could use the following information:

Habitat Information Sheet - Forest

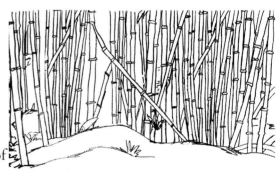

HABITAT: Bamboo Forest

LOCATION AND LOCAL EXPERT: Sichuan, China and Tashi Yidro

AVERAGE RAINFALL

AVERAGE TEMPERATURE

MAIN CHARACTERISTICS: The bamboo forest has lots of quickly growing bamboo. Bamboo is a member of the grass family.

ANIMALS FOUND THERE: Add the panda and eagle (The students could also add red panda).

Animal Record Sheet - Panda

ANIMAL NAME: Panda

CLASSIFICATION: Mammal

FOOD: Herbivore (see note under Food Chart Project in the Additional Activities section)

LOCATION FOUND: Bamboo Forest

INFORMATION LEARNED

- They are endangered animals because their forests are being cut down and they were once hunted for their fur.
- At birth they weigh less than five ounces and have to be carried everywhere.
- Their babies are tiny, blind and pink but after ten weeks they start to crawl.
- They aren't able to walk well until they are about a year old.
- They have usually one to two cubs at a time, which will stay with their mother for about eighteen months.
- Pandas eat bamboo; they have an extra pad on their front paws that works a little like a thumb and helps them to grasp the stems of the bamboo.
- They can eat up to six hundred bamboo stems a day.
- They spend twelve to fifteen hours a day eating.
- They have large powerful jaws for crushing the tough bamboo stems.
- The stems are crushed between their flat cheek teeth and the nutrients are released, but most of the tough bamboo fibers pass straight through their digestive system.
- Their throat has a tough lining to protect them from splinters.

Animal Record Sheet - Eagle

ANIMAL NAME: Golden Eagle

CLASSIFICATION: Bird

FOOD: Carnivore

LOCATION FOUND: Bamboo Forest

INFORMATION LEARNED

- The golden eagle is one of the most widespread eagle species.
- It can be found in Europe and Asia and is named for its golden collar of feathers.
- They can have a wingspan of over seven feet.
- Their outer feathers are very strong and powerful.
- Their inner feathers help to reduce air resistance and give them the ability to soar.
- Their soft down feathers trap air to keep the eagle warm.
- Their eyes face forward so that they can judge distances accurately.
- Their eyes also give them the ability to spot their prey from long distances.
- Eagles have powerful talons and large hooked bills, making them very good hunters.
- They use their talons for killing prey.
- They mate for life and return to the same nest each year.
- Their courtship consists of an intricate flying dance that they perform in pairs.
- They nest in cliffs or trees, they lay one to three eggs per year which hatch in forty-five days.

VOCABULARY

Have your older students look up the following term in the glossary in the appendix on pp. 145-146 or in a science encyclopedia. Have them copy the definition onto a blank index card or into their SCIDAT logbook.

↪ BIRD – A warm-blooded, egg-laying, feathered vertebrate; it also has wings.

(OPTIONAL) COPYWORK

Copywork Sentence

Birds are covered with feathers. They lay eggs in a nest.

Dictation Selection

Birds are warm-blooded animals that are covered with feathers and have wings. They do not give birth to live young. Instead, the mother lays a clutch, or set of eggs. After a period of incubation, the baby birds hatch out of the eggs.

DO: PLAYING WITH SCIENCE

SCIENTIFIC DEMONSTRATION: BIRD BEAKS

Materials
- ☑ Chopsticks
- ☑ Tweezers
- ☑ Pliers
- ☑ Eye dropper
- ☑ Sugar water or honey
- ☑ Gummy worms
- ☑ Unshelled peanuts
- ☑ Seeds
- ☑ Raisins
- ☑ Plate

Procedure
1. Prepare a plate with each of the different types of food on it.
2. Use the chopsticks to try to pick up each of the different types of food and rate the ease of pick up from 1 to 10.
3. Repeat step 2 using the tweezers, pliers, eye dropper.

Explanation

The students should see that it was easier to use:

- The chopsticks to eat the bread (possibly the gummy worms);
- The tweezers to eat the seeds and raisins (possibly the gummy worms);
- The pliers to eat the peanuts;
- The eye dropper to eat the honey.

The chopsticks in this demonstration represent a long, large bill, like that of a stork, whose diet is made up of larger pieces of food such as small fish. In this demonstration, the tweezers represent a short, light beak, like that of a finch, whose diet is made up of small seeds, berries and the occasional worm. In this demonstration, the pliers represent a short, strong beak, like that of a parrot, which uses their strong beak to crack seeds and nuts. In this demonstration, the eye dropper represents a long, slender bill, like that of a hummingbird, whose diet is made up of nectar from flowers.

Take It Further

Have the students repeat the demonstration with different types of food that the students choose.

(Optional) STEAM Projects

Multi-chapter Activities

✂ FOOD CHART – This week, add the panda to the herbivore side and the eagle to the carnivore side of your food chart. You can use the mini-animal pictures found in the appendix on p. 129 of this guide or print out your own. NOTE—There is some debate about whether or not a panda is an herbivore. Although their diet consists only of plant matter, their digestive systems are more similar to that of an omnivore or carnivore. For the sake of simplicity, we have defined the panda as an herbivore, but if you have older students, you may want to have them research this topic further. You can do so using the following blog post:

🖱 https://elementalscience.com/blogs/news/are-giant-pandas-herbivores-or-carnivores

✂ HABITAT PROJECT – Make a poster or diorama that depicts the bamboo forest. This week, add the panda and eagle. You can use the mini-animal pictures found in the appendix on p. 129 of this guide and the habitat poster on p. 135 or print out your own.

Activities For This Chapter

✂ MAKE A PANDA – Make a felt panda face using one large white felt circle, two medium sized black felt circles, two small black felt circles, one black felt triangle and two googly eyes. Use the large white felt circle for the base of the panda face and glue the two medium black felt circles at the top for ears. Then, glue the two small black felt circles in the center top for the eyes of the panda. Next, glue the triangle just below the eyes for the nose. Use a black permanent marker to draw a smile on your panda face and then draw a line from the bottom of the nose to the center of the smile. Finally, glue the two googly eyes onto the two small black felt circles. (NOTE—If you do this project with older students, have them sew on all the pieces as well as the lines for the mouth.)

✂ MAKE AN EAGLE – Make an eagle using the students' hands and feet. Paint one foot golden brown, excluding the toes, and use it to make a print in the center of a sheet of paper. Then, paint the students' left hand golden brown and use it to make a print on the right side of the foot print (fingers will be pointing down). Repeat with the students' right hand on the left side. Add two googly eyes and some white feathers made of paper for the eagle's head. Finally, cut out a beak from yellow construction paper and glue it on.

CHAPTER 12: GRID SCHEDULE

Supplies Needed	
Demo	• Owl Pellet Dissection Kit
Projects	• Bird feeder kit or a soda bottle, Ingredients for deer cookies (Nutter Butter cookie, pretzel, M&M, and frosting)

Chapter Summary

The chapter begins with Willy and Blaine making contact with Ralphy Dingo after Willy shares how the Dingo family should be recognized as noble. Ralphy confirms the presence of another hermit in the Brown Mountains known as Matty Mingo and he warns the two never to go near him. As the three are heading off to bed, Blaine is frightened by an owl, which Willy happily shares more about. Early the next morning Willy wakes up Blaine to go search for the home of the other Feuding Brown Mountain Hermit, and they quickly fall into several of Matty's traps. They end up in an inescapable pit along with a sambar deer which Willy shares about. Ralphy shows up to rescue them, but Matty appears and threatens them all with an explosion. Willy is able to diffuse the situation and the chapter ends with Ralphy and Matty calling a truce.

Weekly Schedule

	Day 1	Day 2	Day 3	Day 4
Read	☐ Read the section entitled "Petrified by the Powerful Owl" of Chapter 12 in *SSA Volume 1: Zoology.*	☐ Read the section entitled "Snared with the Sambar Deer" of Chapter 12 in *SSA Volume 1: Zoology.*	☐ (*Optional*) Read one or all of the assigned pages from the encyclopedia of your choice.	☐ (*Optional*) Read one of the additional library books.
Write	☐ Fill out the Animal Record Sheet on SL p. 51 for the owl. ☐ Go over the vocabulary words and enter them into the Zoology Glossary on SL p. 97.	☐ Fill out the Animal Record Sheet on SL p. 52 for the sambar deer. ☐ Add to the Habitat Info Sheet on SL p. 47 for the Eucalyptus Forest.	☐ Write the information learned from the demonstration on SL p. 54. ☐ (*Optional*) Write a narration on the Zoology Notes Sheet on SL p. 54.	☐ (*Optional*) Complete the copywork or dictation assignment and add it to the Zoology Notes sheet on SL p. 54. ☐ (*Optional*) Take Zoology Quiz #5.
Do	☐ (*Optional*) Make a Bird Feeder.	☐ (*Optional*) Make a Deer Cookie.	☐ Do the demonstration entitled "Owl Pellet Dissection."	☐ (*Optional*) Add the animals studied this week to the food chart and habitat posters.

CHAPTER 12: LIST SCHEDULE

CHAPTER SUMMARY

The chapter begins with Willy and Blaine making contact with Ralphy Dingo after Willy shares how the Dingo family should be recognized as noble. Ralphy confirms the presence of another hermit in the Brown Mountains known as Matty Mingo and he warns the two never to go near him. As the three are heading off to bed, Blaine is frightened by an owl, which Willy happily shares more about. Early the next morning Willy wakes up Blaine to go search for the home of the other Feuding Brown Mountain Hermit, and they quickly fall into several of Matty's traps. They end up in an inescapable pit along with a sambar deer which Willy shares about. Ralphy shows up to rescue them, but Matty appears and threatens them all with an explosion. Willy is able to diffuse the situation and the chapter ends with Ralphy and Matty calling a truce.

ESSENTIALS

Read

☐ Read the section entitled "Petrified by the Powerful Owl" of Chapter 12 in *SSA Volume 1: Zoology*.

☐ Read the section entitled "Snared with the Sambar Deer" of Chapter 12 in *SSA Volume 1: Zoology*.

Write

☐ Fill out the Animal Record Sheet on SL p. 51 for the owl.

☐ Go over the vocabulary words and enter them into the Zoology Glossary on SL p. 97.

☐ Fill out the Animal Record Sheet on SL p. 52 for the sambar deer.

☐ Add to the Habitat Info Sheet on SL p. 47 for the Eucalyptus Forest.

☐ Write the information learned from the demonstration on SL p. 54.

Do

☐ Do the demonstration entitled "Owl Pellet Dissection."

(OPTIONAL) EXTRAS

Read

☐ Read one of the additional library books.

☐ Read one or all of the assigned pages from the encyclopedia of your choice.

Write

☐ Write a narration on the Zoology Notes Sheet on SL p. 54.

☐ Complete the copywork or dictation assignment and add it to the Zoology Notes sheet on SL p. 54.

☐ Take Zoology Quiz #5.

Do

☐ Make a Bird Feeder.

☐ Make a Deer Cookie.

☐ Add the animals studied this week to the food chart and habitat posters.

Supplies Needed	
Demo	• Owl Pellet Dissection Kit
Projects	• Bird feeder kit or a soda bottle, Ingredients for deer cookies (Nutter Butter cookie, pretzel, M&M, and frosting)

Chapter 12: The Feuding Brown Mountain Hermits

Read: Gathering Information

Living Book Reading Assignment

📖 Chapter 12 of *The Sassafras Science Adventures Volume 1: Zoology*

(Optional) Encyclopedia Readings

♀ *Kingfisher First Encyclopedia of Animals* p. 87 (Owl), p. 38 (Deer)

♀ *DK Encyclopedia of Animals* pp. 267-269 (Owl), pp. 154-156 (Deer), pp. 60-61 (Forest)

(Optional) Additional Library Books

📖 *Deer* (Animals That Live in the Forest) by JoAnn Early Macken
📖 *Deer* (Blastoff! Readers: Backyard Wildlife) by Derek Zobel
📖 *The Barn Owl* (Animal Lives) by Bert Kitchen
📖 *White Owl, Barn Owl* by Michael Foreman and Nicola Davies
📖 *There's an Owl in the Shower* by Jean Craighead George
📖 *Forest Bright, Forest Night* (Sharing Nature with Children Books) by Jennifer Ward and Jamichael Henterly

Write: Keeping a Notebook

SCIDAT Logbook Sheets

This week, you can have the students fill out a logbook page for the owl and the deer, as well as continue to fill out a Habitat Information Sheet and the Around the World Sheet for the Eucalyptus Forest. The students could use the following information:

Habitat Information Sheet - Forest

HABITAT: Brown Mountain Forest (or Eucalyptus Forest)

LOCATION AND LOCAL EXPERT

AVERAGE RAINFALL

AVERAGE TEMPERATURE

MAIN CHARACTERISTICS

ANIMALS FOUND THERE: Add the owl and deer.

Around the World - Forest

MAP: Have the students put a star in the top of Victoria, where the Eucalyptus Forest is and label it "Victoria, Australia". Then, have them color the regions of the world that are covered with the eucalyptus forest brown. Use the map pictured as a guide.

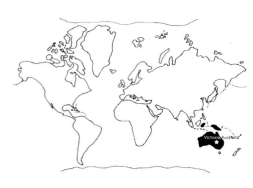

CONTINENTS FOUND: Australia or Oceania

Animal Record Sheet - Owl

ANIMAL NAME: Powerful Owl

CLASSIFICATION: Bird

FOOD: Carnivore

LOCATION FOUND: Brown Mountain Forest

INFORMATION LEARNED

- It is also known as the powerful bookbok.
- They are part of the hawk/owl group of birds.
- They are covered with feathers that are especially soft, which allows them to fly silently.
- They have very sensitive hearing, which allows them to be able to hunt at night even on a moonless night.
- The shape of their head is such that they hear sound on one side a fraction of a second sooner than on the other side, which allows them to pinpoint the location of the sound.
- They have large eyes, which give them good night vision.
- They cannot move their eyes in their sockets; instead, they move their head.
- They swoop down with the feet out and talons spread to catch a prey, usually killing it by piercing it with their talons.
- They mainly catch mice and rabbits.
- Baby owls are called chicks and are covered with soft downy feathers that serve to keep them warm.
- The owl doesn't wait for all its eggs to be laid before incubating them.
- They begin incubation immediately, meaning that their chicks will hatch several days apart.

Animal Record Sheet - Deer

ANIMAL NAME: Sambar Deer

CLASSIFICATION: Mammal

FOOD: Herbivore

LOCATION FOUND: Brown Mountain Forest

INFORMATION LEARNED

- They belong to the family of even toed mammals, including cows, pigs, and camels.
- They live in woodland habitats and use the trees and undergrowth for shelter, food, and protection.
- Their brown color helps them blend into their surroundings.
- Their young are typically covered with spots for camouflage.
- The babies hide in the brush for the first several weeks of their lives, coming out only to feed.
- Each year male deer grow a new set of antlers, adding one point for each year it grows.
- During breeding season they use the antlers to fight with each other to determine who the herd's leader will be.
- The sambar deer have slender, muscular bodies with long thin legs, making them graceful and swift runners, which helps them to escape predators.
- They are plant-eaters, preferring to graze on grass or the occasional leaves, twigs, shoots, flowers, or fruits.

VOCABULARY

Have your older students look up the following term in the glossary in the appendix on pp. 145-146 or in a science encyclopedia. Have them copy the definition onto a blank index card or into their SCIDAT logbook.

☞ FOREST – A habitat that is characterized by the abundance of trees, either deciduous or evergreen.

(OPTIONAL) COPYWORK

Copywork Sentence

A forest is a habitat with lots of trees.

Dictation Selection

Forests are found on one-third of the Earth's land surface. They are characterized by an abundance of trees, which can be deciduous or evergreen. Every forest has three levels: the forest floor, the understory, and the canopy.

(OPTIONAL) QUIZ

This week, you can give the students a quiz based on what they learned in chapters 10 and 12. You can find the quiz in the appendix on p. 163.

Quiz #5 Answers

1. C
2. B
3. D
4. B
5. C

DO: PLAYING WITH SCIENCE

SCIENTIFIC DEMONSTRATION: OWL PELLET

Materials
☑ Owl Pellet Dissection Kit

Procedure

1. Perform an owl pellet dissection using a kit from a science supply store or do virtual dissection at the following website:
   ☞ http://kidwings.com/nests-of-knowledge/virtual-pellet/

   You can see directions and download a printable for this demonstration here:
   ☞ https://elementalscience.com/blogs/science-activities/owl-pellet-dissection

Take It Further

Have the students learn about the common bird calls in your area. You can read *The Little Book of Backyard Bird Songs* by Andrea Pinnington and Caz Buckingham, or you can download the Merlin Bird ID app by Cornell Lab. Have the students sit outside and listen for bird calls, then use what they have learned or use the app to identify the birds they are hearing.

(Optional) STEAM Projects

Multi-chapter Activities

✂ FOOD CHART – This week, add the owl to the carnivore side and deer to the herbivore side of your food chart. You can use the mini-animal pictures found in the appendix on p. 129 of this guide or print out your own.

✂ HABITAT PROJECT – This week, add the owl and deer to your forest habitat. You can use the mini-animal pictures found in the appendix on p. 129 of this guide or print out your own.

Activities For This Chapter

✂ MAKE A BIRD FEEDER – Use a kit to make a wooden bird feeder or make your own using a soda bottle. You can use the directions from the following website:

🖱 http://www.a-home-for-wild-birds.com/soda-bottle-bird-feeder.html

✂ DEER COOKIES – Make some sambar deer cookies using Nutter Butter cookies, pretzels, M&M's and frosting. Break the pretzels in half and use the frosting to attach them at the top of the cookie. Then attach a brown M&M with frosting for the nose and two yellow M&M's for the eyes of the sambar deer. Eat and enjoy!

CHAPTER 13: GRID SCHEDULE

Supplies Needed	
Demo	• No Supplies Needed
Projects	• Popsicle stick, Grey pipe cleaners, Black felt, Black beads

Chapter Summary

The chapter begins with Tashi and Tracey running from a landslide. Once they escape, they meet up with Tashi's sister, Llamo, and begin the trek to her village. On their way, they stop to get Tashi's brother from the monastery where they are greeted by monkeys. Tashi shares quite a bit about the monkeys before the group of four head down the mountain into Tashi's village, where Tracey is welcomed into their home. While waiting for dinner to be ready, Tracey learns about mice from Tashi. After dinner and a long chat, Tashi and Tracey head up to bed. The chapter ends with Tracy snuggling down for the night, only to be awakened by another loud noise!

Weekly Schedule

	Day 1	Day 2	Day 3	Day 4
Read	☐ Read the section entitled "Meeting Monkeys" of Chapter 13 in *SSA Volume 1: Zoology*.	☐ Read the section entitled "Moving in with the Mice" of Chapter 13 in *SSA Volume 1: Zoology*.	☐ (*Optional*) Read one or all of the assigned pages from the encyclopedia of your choice.	☐ (*Optional*) Read one of the additional library books.
Write	☐ Fill out the Animal Record Sheet on SL p. 61 for the monkey. ☐ Go over the vocabulary words and enter them into the Zoology Glossary on SL p. 97.	☐ Fill out the Animal Record Sheet on SL p. 62 for the mouse. ☐ Add to the Habitat Info Sheet on SL p. 57 for the Bamboo Forest.	☐ Write the information learned from the demonstration on SL p. 64. ☐ (*Optional*) Write a narration on the Zoology Notes Sheet on SL p. 64	☐ (*Optional*) Complete the copywork or dictation assignment and add it to the Zoology Notes sheet on SL p. 64. ☐ (*Optional*) Take Zoology Quiz #6.
Do	☐ (*Optional*) Play Monkey See, Monkey Do.	☐ (*Optional*) Make a Mouse Puppet.	☐ Do the demonstration entitled "Primate Eyes."	☐ (*Optional*) Add the animals studied this week to the food chart and habitat posters.

CHAPTER 13: LIST SCHEDULE

CHAPTER SUMMARY

The chapter begins with Tashi and Tracey running from a landslide. Once they escape, they meet up with Tashi's sister, Llamo, and begin the trek to her village. On their way, they stop to get Tashi's brother from the monastery where they are greeted by monkeys. Tashi shares quite a bit about the monkeys before the group of four head down the mountain into Tashi's village, where Tracey is welcomed into their home. While waiting for dinner to be ready, Tracey learns about mice from Tashi. After dinner and a long chat, Tashi and Tracey head up to bed. The chapter ends with Tracy snuggling down for the night, only to be awakened by another loud noise!

ESSENTIALS

Read

- ☐ Read the section entitled "Meeting Monkeys" of Chapter 13 in *SSA Volume 1: Zoology*.
- ☐ Read the section entitled "Moving in with the Mice" of Chapter 13 in *SSA Volume 1: Zoology*.

Write

- ☐ Fill out the Animal Record Sheet on SL p. 61 for the monkey.
- ☐ Go over the vocabulary words and enter them into the Zoology Glossary on SL p. 97.
- ☐ Fill out the Animal Record Sheet on SL p. 62 for the mouse.
- ☐ Add to the Habitat Info Sheet on SL p. 57 for the Bamboo Forest.
- ☐ Write the information learned from the demonstration on SL p. 64.

Do

- ☐ Do the demonstration entitled "Primate Eyes."

(OPTIONAL) EXTRAS

Read

- ☐ Read one of the additional library books.
- ☐ Read one or all of the assigned pages from the encyclopedia of your choice.

Write

- ☐ Add to the Habitat Info Sheet on SL p. 57 for the Bamboo Forest.
- ☐ Write a narration on the Zoology Notes Sheet on SL p. 64.
- ☐ Complete the copywork or dictation assignment and add it to the Zoology Notes sheet on SL p. 64.
- ☐ Take Zoology Quiz #6.

Do

- ☐ Play Monkey See, Monkey Do.
- ☐ Make a Mouse Puppet.
- ☐ Add the animals studied this week to the food chart and habitat posters.

Supplies Needed	
Demo	• No Supplies Needed
Projects	• No Additional Supplies Needed

Chapter 13: Trekking in Sichuan

Read: Gathering Information

Living Book Reading Assignment

📖 Chapter 13 of *The Sassafras Science Adventures Volume 1: Zoology*

(Optional) Encyclopedia Readings

🔖 *Kingfisher First Encyclopedia of Animals* p. 47 (Monkey), p. 60 (Mouse)

🔖 *DK Encyclopedia of Animals* pp. 252-254 (Monkey), pp. 246-247 (Mouse)

(Optional) Additional Library Books

📖 *Monkeys* (Animals That Live in the Rain Forest) by Julie Guidone

📖 *Monkeys and Apes* (Read About Animals) by Dean Morris

📖 *If You Give a Mouse a Cookie* (If You Give...) by Laura Joffe Numeroff and Felicia Bond

📖 *The Life Cycle of a Mouse* (The Life Cycles Library) by Andrew Hipp

Write: Keeping a Notebook

SCIDAT Logbook Sheets

This week, you can have the students fill out a logbook page for the monkey and the mouse, as well as continue to fill out a Habitat Information Sheet and the Around the World Sheet for the Bamboo Forest. The students could use the following information:

Habitat Information Sheet

HABITAT: Bamboo Forest

LOCATION AND LOCAL EXPERT

AVERAGE RAINFALL

AVERAGE TEMPERATURE

MAIN CHARACTERISTICS

ANIMALS FOUND THERE: Add the monkey and mouse.

Around the World

MAP: Have the students put a star on China, where the Bamboo Forest is and label it "Sichuan, China". Then, have them color the regions of the world that are covered with the bamboo forest habitat purple. Use the map pictured as a guide.

CONTINENTS FOUND: Asia, Africa, Australia or Oceania, North America

Animal Record Sheet - Monkey

ANIMAL NAME: Golden Hair Monkey

CLASSIFICATION: Mammal

FOOD: Herbivore

LOCATION FOUND: China

INFORMATION LEARNED

- They are from the primate family.
- They got their name from the golden hair all around their head.
- Their fur is a bit thicker allowing them to live in an area that is a bit colder.
- They are very intelligent animals.
- They can solve problems and hold things in their hands.
- Their strong flexible hands are perfect for grasping branches.
- They are good climbers.
- They have long tails that help them balance.
- They live in troops of up to two hundred animals in the summer, but only twenty to thirty animals in the winter.
- Males can be up to twenty pounds, while females are closer to thirteen pounds; babies are usually born from March to June after a six month gestation.
- Their primary source of food is lichens, but they will also eat tree bark, leaves, flowers, and fruit.

Animal Record Sheet - Mouse

ANIMAL NAME: Mouse

CLASSIFICATION: Mammal

FOOD: Herbivore

LOCATION FOUND: China

INFORMATION LEARNED

- A female can have ten liters per year, with up to twelve babies in each litter.
- Baby mice are born naked, blind, and without ears, so they are completely helpless for the first few days.
- By two weeks the baby mice can leave the nest.
- Found all over the world, they are the most numerous and diverse group of mammals.
- They are small rodents with long tails.
- They eat seeds, grain, fruit, insects, and sometimes human food.
- They have a good sense of sight, hearing and smell as well as highly sensitive whiskers for finding their way in the dark.
- They have sharp front teeth that constantly grow, so they must continually gnaw at things to keep them from getting too long.

VOCABULARY

Have your older students look up the following term in the glossary in the appendix on pp. 145-146 or in a science encyclopedia. Have them copy the definition onto a blank index card or into their SCIDAT logbook.

⊘ VERTEBRATE – An animal that has an internal skeleton made from bone.

(OPTIONAL) COPYWORK

Copywork Sentence

A vertebrate is an animal with a backbone.

Dictation Selection

Vertebrates are animals that have an internal skeleton made out of bone. They include birds, fish, amphibians, reptiles, and mammals. Monkeys, mice, eagles, and pandas are all examples of vertebrates.

(OPTIONAL) QUIZ

This week, you can give the students a quiz based on what they learned in chapters 11 and 13. You can find the quiz in the appendix on p. 165.

Quiz #6 Answers
1. D
2. A
3. D
4. C
5. D

DO: PLAYING WITH SCIENCE

SCIENTIFIC DEMONSTRATION: PRIMATE EYES

Materials
☑ The student

Procedure
1. Have the students stand tall in the middle of a room and extend both arms out to their side.
2. Then, have them touch their index finger from one hand to their nose. Repeat with the index finger from the other hand.
3. Next have them repeat step 2 with one eye closed. Ask the students:

? Did you see any difference?

Explanation

The students should have seen that it was much easier to bring their finger to their nose when both eyes were open. Animals have different eye positions that give them the field of vision they need to carry out their daily tasks. Primates and humans have eyes on the front of their head that can view things from a slightly different angle, giving them stereoscopic vision. Their brain then merges the images to give them a 3-D picture of what they are seeing, which helps these animals to be able to judge distances. This is why it is easier for a human to see with both eyes open. Herbivores generally have eyes that are positioned on the sides of their heads, giving them lateral vision. This helps them to keep an eye out for predators while they are grazing for food. Animals that live in trees or hunt for their food usually have eyes on the front of their head, giving them binocular vision. This allows them to focus on their prey from a long distance.

Take It Further

Read aloud the following book to the students over the next few weeks:

📖 *My Life with the Chimpanzees* by Jane Goodall

(Optional) STEAM Projects

Multi-chapter Activities

✂ FOOD CHART – This week, add the monkey and mouse to the herbivore side of your food chart. You can use the mini-animal pictures found in the appendix on p. 129 of this guide or print out your own.

✂ HABITAT PROJECT – This week, add the monkey and mouse to your bamboo forest habitat. You can use the mini-animal pictures found in the appendix on p. 129 of this guide or print out your own.

Activities For This Chapter

✂ PLAY MONKEY SEE, MONKEY DO – You will need three or more players. Begin with sending one player out of the room. Then, have the remaining players select one to be the monkey. They will make movements and gestures that must be imitated by the other players. Next have the absent player come back in and try to guess which of the players is the monkey. If the monkey gets caught, he becomes the next one to leave the room.

✂ MOUSE PUPPET – You will need a popsicle stick, 2 grey pipe cleaners, a few scraps of pink felt and 2 small black beads. Begin by cutting one of the pipe cleaners in half and then take one of those pieces and cut it in half once more. You will end up with 4 pieces, one large, one medium and two short. Roll all four pieces into a flat spiral, leaving a little loose tail on the larger piece. Then glue the medium spiral (the head) to the larger one (the body), and the two small spirals (the ears) to the top of the medium spiral. Cut out three scraps of pink felt, one for the nose and two for the ears. Attach them to your mouse, glue the black beads on for eyes and attach your mouse to a popsicle stick.

CHAPTER 14: GRID SCHEDULE

Supplies Needed	
Demo	• 2 Glass jars, Box at least 2 inches wider and taller than the jars, Cotton balls, 2 Thermometers
Projects	• Decorative gourd, Leaves and twigs, Cloves

Chapter Summary

The chapter begins with Blaine and Tracey reuniting in the underground laboratory of Summer Beach, who has been expecting them. She introduces them to her lab assistant Ulysses S. Grant, the arctic ground squirrel, and then explains that she and Uncle Cecil were school-mates. Summer explains that things will be a bit different for this leg of their journey. They will learn all the SCIDAT data in her lab and then they will go out to see the animals in their natural habitat. She shares information on the musk ox and snow goose before letting them rest for the night. In the morning, they take off in Summer's heliquicker to go take pictures of the animals. The chapter ends with the group blissfully unaware of the fact that they are headed right into a storm.

Weekly Schedule

	Day 1	Day 2	Day 3	Day 4
Read	☐ Read the section entitled "Arctic Reunion" of Chapter 14 in *SSA Volume 1: Zoology*.	☐ Read the section entitled "Animals in a Flash" of Chapter 14 in *SSA Volume 1: Zoology*.	☐ (*Optional*) Read one or all of the assigned pages from the encyclopedia of your choice.	☐ (*Optional*) Read one of the additional library books.
Write	☐ Go over the vocabulary words and enter them into the Zoology Glossary on SL p. 97. ☐ Add to the Habitat Information Sheet for the Arctic on SL p. 67.	☐ Fill out the Animal Record Sheet for the musk ox on SL p. 69 and for the snow goose on SL p. 70.	☐ Write the information learned from the demonstration on SL p. 73. ☐ (*Optional*) Write a narration on the Zoology Notes Sheet on SL p. 73.	☐ (*Optional*) Complete the copywork or dictation assignment and add it to the Zoology Notes sheet on SL p. 73.
Do	☐ (*Optional*) Compare the Tundra versus Taiga.	☐ (*Optional*) Make a Squash Goose.	☐ Do the demonstration entitled "Hairy Fur."	☐ (*Optional*) Add the animals studied this week to the food chart and habitat posters.

CHAPTER 14: LIST SCHEDULE

CHAPTER SUMMARY

The chapter begins with Blaine and Tracey reuniting in the underground laboratory of Summer Beach, who has been expecting them. She introduces them to her lab assistant Ulysses S. Grant, the arctic ground squirrel, and then explains that she and Uncle Cecil were school-mates. Summer explains that things will be a bit different for this leg of their journey. They will learn all the SCIDAT data in her lab and then they will go out to see the animals in their natural habitat. She shares information on the musk ox and snow goose before letting them rest for the night. In the morning, they take off in Summer's heliquicker to go take pictures of the animals. The chapter ends with the group blissfully unaware of the fact that they are headed right into a storm.

ESSENTIALS

Read

☐ Read the section entitled "Arctic Reunion" of Chapter 14 in *SSA Volume 1: Zoology.*

☐ Read the section entitled "Animals in a Flash" of Chapter 14 in *SSA Volume 1: Zoology.*

Write

☐ Go over the vocabulary words and enter them into the Zoology Glossary on SL p. 97.

☐ Add to the Habitat Information Sheet for the Arctic on SL p. 67.

    ☐ Fill out the Animal Record Sheet for the musk ox on SL p. 69 and for the snow goose on SL p. 70.

    ☐ Write the information learned from the demonstration on SL p. 73.

Do

☐ Do the demonstration entitled "Hairy Fur."

(OPTIONAL) EXTRAS

Read

    ☐ Read one of the additional library books.

    ☐ Read one or all of the assigned pages from the encyclopedia of your choice.

Write

    ☐ Write a narration on the Zoology Notes Sheet on SL p. 73.

    ☐ Complete the copywork or dictation assignment and add it to the Zoology Notes sheet on SL p. 73.

Do

    ☐ Compare the Tundra versus Taiga.

    ☐ Make a Squash Goose.

    ☐ Add the animals studied this week to the food chart and habitat posters.

	Supplies Needed
Demo	• 2 Glass jars, Box at least 2 inches wider and taller than the jars, Cotton balls, 2 Thermometers
Projects	• Decorative gourd, Leaves and twigs, Cloves

Chapter 14: Arctic Adventures

READ: Gathering Information

Living Book Reading Assignment

📖 Chapter 14 of *The Sassafras Science Adventures Volume 1: Zoology*

(Optional) Encyclopedia Readings

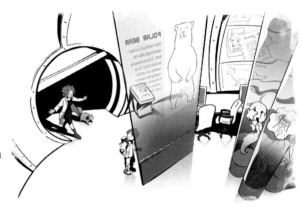

🔖 *Kingfisher First Encyclopedia of Animals* p. 95 (Duck and Goose)

🔖 *DK Encyclopedia of Animals* p. 241 (paragraph on bottom half), p. 184 (Geese), pp. 54-57 (Arctic)

(Optional) Additional Library Books

📖 *Musk Oxen* (Animals That Live in the Tundra) by Roman Patrick

📖 *The Itchy Little Musk Ox* by Tricia Brown and Debra Dubac

📖 *Honk, Honk, Goose!: Canada Geese Start a Family* by April Pulley Sayre and Huy Voun Lee

📖 *The Goose Man: The Story of Konrad Lorenz* by Elaine Greenstein

📖 *Over in the Arctic: Where the Cold Winds Blow* (Sharing Nature with Children Books) by Marianne Berkes and Jill Dubin

📖 *The Arctic Habitat* (Introducing Habitats) by Molly Aloian and Bobbie Kalman

WRITE: Keeping a Notebook

SCIDAT Logbook Sheets

This week, you can have the students begin to fill out a Habitat Information Sheet for the Arctic (Tundra or Taiga) and a logbook page for the musk ox and snow goose. The students could use the following information:

Habitat Information Sheet - Arctic

NOTE—In this leg of the Sassafras twins' journey they are visiting both the Arctic Tundra and Taiga. There are many similarities between the Arctic Tundra and Taiga, but there are a few differences as well. For the sake of simplicity we have chosen to refer to this as the "Arctic Tundra" in the book, as it is the more widely recognized Arctic habitat, and "Arctic" in the student logbook. If you have older students we would recommend that they explore the similarities and differences on a deeper level (see the additional activities for more information). Either way, we recommend that you wait until next week to completely fill out the Habitat Information Sheet.

HABITAT: Arctic (Tundra or Taiga)

LOCATION AND LOCAL EXPERT: Alaska and Summer Beach

AVERAGE RAINFALL

AVERAGE TEMPERATURE

MAIN CHARACTERISTICS

ANIMALS FOUND THERE: Add the musk ox and snow goose.

Animal Record Sheet - Ox

ANIMAL NAME: Musk Ox

CLASSIFICATION: Mammal

FOOD: Herbivore

LOCATION FOUND: Alaska

INFORMATION LEARNED
- They are found in Alaska, northern Canada, and Greenland.
- They were once wiped out of Alaska and north-western Canada in the late 19th century due to over hunting; they were re-introduced into parts of Alaska in 1935 and have since thrived.
- They are related to sheep and goats.
- They feed on grasses, lichens, and moss mainly in the summer and arctic willows in the winter.
- They have a thick, shaggy coat to protect them from the cold.
- The long outer layers of coarse hair protect them from snow and rain, while their thick, dense under fur keeps them warm and dry.
- Their wool is being collected on farms to be used for winter clothing.
- The males emit a heavy odor called musk to attract females for mating and to mark their territory.
- They will also fight other males by charging into each other until one gives up for the right to breed with a certain female.
- Females give birth after eight to nine months gestation period to one calf, which can keep up with the herd after only a few hours of life.
- The calves begin eating grasses after two months.
- The adult oxen will gather together in a defensive huddle when threatened, every adult standing with its back to the center of the huddle and its horns facing towards the attacker.

Animal Record Sheet - Goose

ANIMAL NAME: Snow Goose

CLASSIFICATION: Bird

FOOD: Herbivore

LOCATION FOUND: Alaska

INFORMATION LEARNED
- Geese belong to the family of water birds known as waterfowl, but they spend more time on land than ducks and swans do.
- They have webbed feet and boat-shaped bodies that make them able to swim well when they need to.
- Geese travel in large family groups called flocks.
- They travel together to look for warm weather and food and warn each other of predators with a loud honking call.
- They migrate from the Gulf of Mexico to the Arctic every summer.
- Geese are generally larger than ducks and have a much longer neck than ducks do.
- They eat grass, marsh plants, and other grains.

- They use their broad bill to grip and rip grasses and plants.
- They also have a sharp "tooth" on the side of their beak that they use for tearing the toughest stalks.
- The mother goose builds her nest on an area of high ground, using plant matter and feathers.
- She will lay three to five eggs and then incubate them for twenty-two to twenty-five days before they hatch.
- Baby geese are called goslings; within a few hours of birth the goslings can swim, walk, and find food.

VOCABULARY

Have your older students look up the following term in the glossary in the appendix on pp. 145-146 or in a science encyclopedia. Have them copy the definition onto a blank index card or into their SCIDAT logbook.

✎ ARCTIC – A habitat that has little vegetation and very cold temperatures.

(OPTIONAL) COPYWORK

Copywork Sentence

The arctic has very few plants. It is also very cold.

Dictation Selection

The arctic tundra has a permanently frozen subsoil that makes it impossible for trees to grow. The plants that are typically found there are mosses and lichens. The tundra has very long and cold winters. Even so, a few animals, like the polar bear and mountain goat, have adapted to life there.

DO: PLAYING WITH SCIENCE

SCIENTIFIC DEMONSTRATION: HAIRY FUR

Materials
- ☑ 2 Glass jars
- ☑ Box at least 2 inches wider and taller than the jars
- ☑ Cotton balls
- ☑ 2 Thermometers

Procedure
1. Fill two glasses with hot water and record the temperature of each.
2. Leave one glass on the counter. Place the other glass in the box and quickly pack the cotton balls around it, trying to fill up all the space between the glass and the box.
3. Let the glasses sit for 15 minutes and record the temperatures again.

Explanation

The students should see that the temperature of the glass that sat out on the counter should be lower. The cotton in the box helped to insulate the water in the glass and prevent it from losing its heat too quickly. The hair on a mammal also acts as an insulator and helps to keep their body temperature from dropping too quickly.

Take It Further

Have the students repeat the demonstration using feathers and wool.

(OPTIONAL) STEAM PROJECTS

Multi-chapter Activities

✄ FOOD CHART – This week, add the musk ox and snow goose to the herbivore side of your food chart. You can use the mini-animal pictures found in the appendix on p. 129 of this guide or print out your own.

✄ HABITAT PROJECT – Make a poster or diorama that depicts the arctic. This week, add the musk ox and snow goose. You can use the mini-animal pictures found in the appendix on p. 129 of this guide and the habitat poster on p. 136 or print out your own.

Activities For This Chapter

✄ TUNDRA VERSUS TAIGA – Have the students create a chart showing the similarities and differences between the arctic tundra and taiga. The most notable similarities are the harsh climate, the fact that both animals and plants have adapted to live there, and that they are both found in the outermost regions of our globe. The most notable differences are that the Taiga has coniferous trees, while the tundra only has mosses and lichens, the tundra has a permanently frozen layer that prevents trees from growing, and the tundra tends to be found further north than the Taiga.

✄ SQUASH GOOSE – Have the students select a squash or decorative gourd with a crooked neck. Then glue on leaves for wings and rocks or twigs for feet. Use cloves for eyes and then carve the stem of the squash to look like a beak.

CHAPTER 15: GRID SCHEDULE

Supplies Needed	
Demo	• 1 Large plastic bag, Stopwatch, Rubber band, Plastic glove, Shortening, Tub of ice water
Projects	• Chalk pastels, Blue construction paper, Ingredients for cheese (goat milk, vinegar, and salt), Cheesecloth, Thermometer, Pot

Chapter Summary

The chapter opens with a dilemma, Uncle Cecil must risk going into Old Man Grusher's backyard, possibly facing off with his Guardian Beast, in order to retrieve the petri dish, otherwise known as a frisbee, that landed there during a game of Pass the Petri. After several pep talks, he finally makes it over the fence with the help of Blaine and Tracey. They meet Old Man Grusher and learn that he is not as bad as they thought before heading back to Cecil's basement lab where the twins get to hear President Lincoln's ever-so-brief presentation on geology. We learn that the next leg of the twin's journey is Zoology and Summer is going to be their local expert. Before the chapter closes, we also learn that the Man With No Eyebrows has not given up, in fact he has a whole army of scientists helping him now, thanks to Adrienne Archer, the rough Swiss Secret Service agent!

Weekly Schedule

	Day 1	Day 2	Day 3	Day 4
Read	☐ Read the section entitled "Summer Storms" of Chapter 15 in *SSA Volume 1: Zoology.*	☐ Read the section entitled "Searching for Sassafras" of Chapter 15 in *SSA Volume 1: Zoology.*	☐ (*Optional*) Read one or all of the assigned pages from the encyclopedia of your choice.	☐ (*Optional*) Read one of the additional library books.
Write	☐ Add to the Habitat Information Sheet for the Arctic on SL p. 67. ☐ Go over the vocabulary words and enter them into the Zoology Glossary on SL p. 98.	☐ Fill out the Animal Record Sheet for the polar bear on SL p. 71 and for the mountain goat on SL p. 72.	☐ Write the information learned from the demonstration on SL p. 74. ☐ (*Optional*) Write a narration on the Zoology Notes Sheet on SL p. 74.	☐ (*Optional*) Complete the copywork or dictation assignment and add it to the Zoology Notes sheet on SL p. 74. ☐ (*Optional*) Take Zoology Quiz #7.
Do	☐ (*Optional*) Make Polar Bear Art.	☐ (*Optional*) Make Goat Cheese.	☐ Do the demonstration entitled "Blubber."	☐ (*Optional*) Add the animals studied this week to the food chart and habitat posters.

Chapter 15: List schedule

Chapter Summary

The chapter opens with a dilemma, Uncle Cecil must risk going into Old Man Grusher's backyard, possibly facing off with his Guardian Beast, in order to retrieve the petri dish, otherwise known as a frisbee, that landed there during a game of Pass the Petri. After several pep talks, he finally makes it over the fence with the help of Blaine and Tracey. They meet Old Man Grusher and learn that he is not as bad as they thought before heading back to Cecil's basement lab where the twins get to hear President Lincoln's ever-so-brief presentation on geology. We learn that the next leg of the twin's journey is Zoology and Summer is going to be their local expert. Before the chapter closes, we also learn that the Man With No Eyebrows has not given up, in fact he has a whole army of scientists helping him now, thanks to Adrienne Archer, the rough Swiss Secret Service agent!

Essentials

Read

- ☐ Read the section entitled "Summer Storms" of Chapter 15 in *SSA Volume 1: Zoology*.
- ☐ Read the section entitled "Searching for Sassafras" of Chapter 15 in *SSA Volume 1: Zoology*.

Write

- ☐ Add to the Habitat Information Sheet for the Arctic on SL p. 67.
- ☐ Go over the vocabulary words and enter them into the Zoology Glossary on SL p. 98.
- ☐ Fill out the Animal Record Sheet for the polar bear on SL p. 71 and for the mountain goat on SL p. 72.
- ☐ Write the information learned from the demonstration on SL p. 74.

Do

- ☐ Do the demonstration entitled "Blubber."

(Optional) Extras

Read

- ☐ Read one of the additional library books.
- ☐ Read one or all of the assigned pages from the encyclopedia of your choice.

Write

- ☐ Write a narration on the Zoology Notes Sheet on SL p. 74.
- ☐ Complete the copywork or dictation assignment and add it to the Zoology Notes sheet on SL p. 74.
- ☐ Take Zoology Quiz #7.

Do

- ☐ Make Polar Bear Art.
- ☐ Make Goat Cheese.
- ☐ Add the animals studied this week to the food chart and habitat posters.

Supplies Needed	
Demo	• 1 Large plastic bag, Stopwatch, Rubber band, Plastic glove, Shortening, Tub of ice water
Projects	• Chalk pastels, Blue construction paper, Ingredients for cheese (goat milk, vinegar, and salt), Cheesecloth, Thermometer, Pot

Chapter 15: Split Up by the Storm

Read: Gathering Information

Living Book Reading Assignment

📖 Chapter 15 of *The Sassafras Science Adventures Volume 1: Zoology*

(Optional) Encyclopedia Readings

🔖 *Kingfisher First Encyclopedia of Animals* p. 42 (Polar Bear), p. 76 (Goat)

🔖 *DK Encyclopedia of Animals* pp. 283-284 (Polar Bear), p. 191 (Goat)

(Optional) Additional Library Books

📖 *Face to Face with Polar Bears* (Face to Face with Animals) by Norbert Rosing and Elizabeth Carney

📖 *Polar Bears* by Mark Newman

📖 *Where Do Polar Bears Live?* (Let's-Read-and-Find... Science 2) by Sarah L. Thomson

📖 *Goats* (Animals That Live on the Farm) by JoAnn Early Macken

📖 *Life on a Goat Farm* (Life on a Farm) by Judy Wolfman

📖 *Little Apple Goat* by Caroline Church

Write: Keeping a Notebook

SCIDAT Logbook Sheets

This week, you can have the students fill out a logbook page for the polar bear and the mountain goat, as well as continue to fill out a Habitat Information Sheet and the Around the World Sheet for the Arctic. The students could use the following information:

Habitat Information Sheet - Arctic

HABITAT: Arctic (Tundra or Taiga)

LOCATION AND LOCAL EXPERT: Alaska and Summer Beach

AVERAGE RAINFALL: The arctic has lots of snowfall.

AVERAGE TEMPERATURE: It has super cold winters with cool and humid summers.

MAIN CHARACTERISTICS: There are plentiful lichens, coni

ANIMALS FOUND THERE: Add the polar bear and goat.

Around the World - Arctic Habitat

MAP: Have the students put a star in the top of Alaska, where the Arctic habitat is and label it "Alaska". Then, have them color the regions of the world that are covered with the arctic habitat light blue. Use the map pictured as a guide.

CONTINENTS FOUND: North America, Antarctica, Asia

Animal Record Sheet - Bear

ANIMAL NAME: Polar Bear

CLASSIFICATION: Mammal

FOOD: Carnivore

LOCATION FOUND: Alaska

INFORMATION LEARNED

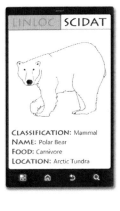

- Polar bears are the world's largest meat-eating land animals.
- They are found only in the arctic regions of Alaska, Canada, Russia, Norway, and Greenland.
- They are excellent hunters, eating mainly seals, but also fish, geese, and ducks.
- Polar bears are excellent swimmers because their large furry feet make good paddles.
- They live mainly on the ice floes, so they occasionally need to swim from one ice patch to another.
- They can swim for up to twenty-five miles.
- Polar bears are the only northern bears that do not all hibernate in the winter.
- Pregnant polar bears do hibernate in dens, in which they give birth.
- Polar bear babies are about the size of a rat when born.
- They weigh less than two pounds at birth, but they grow quickly.
- After several months in the den the babies venture out for the first time with their mother, but they stay with her for up to two years.
- They have a thick, oily coat of white fur and a layer of fat that protects them from temperatures that are below freezing.
- Their hair is hollow so that it can trap heat like a greenhouse.
- The oily substance on their fur makes it waterproof.

Animal Record Sheet - Goat

ANIMAL NAME: Mountain Goat

CLASSIFICATION: Mammal

FOOD: Herbivore

LOCATION FOUND: Alaska

INFORMATION LEARNED

- The mountain goats of Alaska are one of two all-white hoofed species in Alaska.
- Goats belong to the cattle family.
- Males are called billies, females are called nannies, and their young are called kids.
- Goats are hardy climbers that can survive some of the highest mountains.
- They are found all across the Northern hemisphere in cold, mountainous places.
- They have split hooves with hard edges and soft centers that act like suction cups and give them stability on slippery rocks.
- They have a long shaggy winter coat, which keeps out the cold.
- Around June they shed their winter coat and are left with a soft summer coat, which grows into a full winter coat by the time the season comes around again.

- They prefer to feed on grasses and plants, but will eat almost any plant matter to survive, including trees and shrubs.
- Some goats have been domesticated and are used for their wool, milk and meat.

VOCABULARY

Have your older students look up the following term in the glossary in the appendix on pp. 145-146 or in a science encyclopedia. Have them copy the definition onto a blank index card or into their SCIDAT logbook.

↪ HIBERNATION – The ability of an animal to go into a deep sleep for a long period of time by shutting down its body.

(OPTIONAL) COPYWORK

Copywork Sentence

Animals that hibernate fall into a deep sleep in the winter.

Dictation Selection

Hibernation is the ability of an animal to go into a deep sleep for a long period of time. Animals typically hibernate in the winter because the temperature is cold and food is hard to find. The squirrel, the frog, and the black bear are all examples of animals that hibernate during the winter.

(OPTIONAL) QUIZ

This week, you can give the students a quiz based on what they learned in chapters 14 and 15. You can find the quiz in the appendix on p. 167.

Quiz #7 Answers
1. C
2. D
3. B
4. B
5. C

DO: PLAYING WITH SCIENCE

SCIENTIFIC DEMONSTRATION: BLUBBER

Materials
- ☑ 1 Large plastic bag
- ☑ Rubber band
- ☑ Plastic glove
- ☑ Shortening
- ☑ Tub of ice water
- ☑ Stopwatch

Procedure
1. Put the glove on your hand and place your hand in the bucket of ice water. Time how long it is until you pull your hand out.

2. While your hand warms up, fill the plastic bag half way with shortening. Place your gloved hand into the plastic bag and mold the shortening around your hand, so that it is completely coated.
3. While still in the bag, place your hand in the bucket of ice water and time how long it is until you pull your hand out.

Explanation

The students should be able to keep their hand in the ice water much longer the second time. The shortening insulates their hand from the cold temperature in the same what that animal blubber insulates the polar bear. Blubber is a thick layer of fat that can be found in polar bears and other arctic animals. It insulates them from the cold temperatures of the habitat they live in and it serves as an energy store for the mammal.

Take It Further

Have the students repeat the demonstration with warm water to see if the results they get are the same.

(OPTIONAL) STEAM PROJECTS

Multi-chapter Activities

✂ FOOD CHART – This week, add the polar bear to the carnivore side and the mountain goat to the herbivore side of your food chart. You can use the mini-animal pictures found in the appendix on p. 129 of this guide or print out your own.

✂ HABITAT PROJECT – This week, add the polar bear and the mountain goat to your arctic habitat. You can use the mini-animal pictures found in the appendix on p. 129 of this guide or print out your own.

Activities For This Chapter

✂ POLAR BEAR ART – Draw an arctic scene, complete with polar bears. The level of detail for this project will depend on the age of the students. For younger students, consider letting them paint a darkly colored sheet of paper mostly white. Then once it is dry, use a marker to draw the polar bears. For older students, consider having them use pastels to create a winter scene. Then use a very light shade of yellow to draw a polar bear so that it stands out a bit from the background.

✂ GOAT CHEESE – You will need goat milk, vinegar, salt, a thermometer, a non-reactive pot, cheesecloth and a fine mesh colander. (NOTE—If you don't have access to goat's milk, you can substitute it with cow's milk. If you don't have access to cheesecloth, you can substitute it with paper towels.) Heat the milk over medium heat, stirring continuously, until it reaches 180°F to 185°F. You will see small bubbles and the surface will be foamy. Remove from the heat and add one third of a cup of white vinegar, stir to mix and then let the mixture sit for 10 minutes (curds will form). Meanwhile, line the colander with cheesecloth. Pour your curds into the colander and let the whey strain out until you reach the desired consistency of cheese.

Chapter 16: Grid Schedule

	Supplies Needed
Demo	• Old plastic doll with hair, such as a Barbie or pony doll, Tub of water
Projects	• Egg carton, Construction paper, Construction paper, Paper clips, Magnet, Dowel rod, String

Chapter Summary

The chapter opens with the twins' shocking and watery landing. They are rescued by the crew of the *Scot's Folly III*, where they find out that their local expert for this leg, Captain James Q. McScruffy, is not a very open man. The first mate William Atwater does tell them that if they get close enough they can hear the captain mumbling all sorts of facts about animals, or they can just ply him with grape soda. They find out from the captain that the *Scot's Folly III* is on a hunt for treasure that was stolen from his great-grandfather by a giant squid. Later, as they are cleaning the deck, they spot King Penguins on the shore and the twins get up close to Captain McScruffy to hear all about the animals. The *Scot's Folly III* is then boarded by the captain of the P.R.O. pirates, who turns out to just be looking for donations for their organization. As Captain Peach Beard leaves, they spot a school of codfish and the twins listen intently as Captain McScruffy mumbles information about them. The chapter ends with the ship heading due south in search of the treasure.

Weekly Schedule

	Day 1	Day 2	Day 3	Day 4
Read	☐ Read the section entitled "Promising Penguins" of Chapter 16 in *SSA Volume 1: Zoology.*	☐ Read the section entitled "Comical Codfish" of Chapter 16 in *SSA Volume 1: Zoology.*	☐ (*Optional*) Read one or all of the assigned pages from the encyclopedia of your choice.	☐ (*Optional*) Read one of the additional library books.
Write	☐ Fill out the Animal Record Sheet on SL p. 79 for the penguin. ☐ Go over the vocabulary words and enter them into the Zoology Glossary on SL p. 98.	☐ Fill out the Animal Record Sheet on SL p. 80 for the fish. ☐ Add to the Habitat Information Sheet for the Ocean on SL p. 77.	☐ Write the information learned from the demonstration on SL p. 83. ☐ (*Optional*) Write a narration on the Zoology Notes Sheet on SL p. 83.	☐ (*Optional*) Complete the copywork or dictation assignment and add it to the Zoology Notes sheet on SL p. 83.
Do	☐ (*Optional*) Make a Penguin.	☐ (*Optional*) Go Magnetic Fishing.	☐ Do the demonstration entitled "Fish Gills."	☐ (*Optional*) Add the animals studied this week to the food chart and habitat posters.

CHAPTER 16: LIST SCHEDULE

CHAPTER SUMMARY

The chapter opens with the twins' shocking and watery landing. They are rescued by the crew of the *Scot's Folly III*, where they find out that their local expert for this leg, Captain James Q. McScruffy, is not a very open man. The first mate William Atwater does tell them that if they get close enough they can hear the captain mumbling all sorts of facts about animals, or they can just ply him with grape soda. They find out from the captain that the *Scot's Folly III* is on a hunt for treasure that was stolen from his great-grandfather by a giant squid. Later, as they are cleaning the deck, they spot King Penguins on the shore and the twins get up close to Captain McScruffy to hear all about the animals. The *Scot's Folly III* is then boarded by the captain of the P.R.O. pirates, who turns out to just be looking for donations for their organization. As Captain Peach Beard leaves, they spot a school of codfish and the twins listen intently as Captain McScruffy mumbles information about them. The chapter ends with the ship heading due south in search of the treasure.

ESSENTIALS

Read

- ☐ Read the section entitled "Promising Penguins" of Chapter 16 in *SSA Volume 1: Zoology*.
- ☐ Read the section entitled "Comical Codfish" of Chapter 16 in *SSA Volume 1: Zoology*.

Write

- ☐ Fill out the Animal Record Sheet on SL p. 79 for the penguin.
- ☐ Go over the vocabulary words and enter them into the Zoology Glossary on SL p. 98.
- ☐ Fill out the Animal Record Sheet on SL p. 80 for the fish.
- ☐ Add to the Habitat Information Sheet for the Ocean on SL p. 77.
- ☐ Write the information learned from the demonstration on SL p. 83.

Do

- ☐ Do the demonstration entitled "Fish Gills."

(OPTIONAL) EXTRAS

Read

- ☐ Read one of the additional library books.
- ☐ Read one or all of the assigned pages from the encyclopedia of your choice.

Write

- ☐ Write a narration on the Zoology Notes Sheet on SL p. 83.
- ☐ Complete the copywork or dictation assignment and add it to the Zoology Notes sheet on SL p. 83.

Do

- ☐ Make a Penguin.
- ☐ Go Magnetic Fishing.
- ☐ Add the animals studied this week to the food chart and habitat posters.

Supplies Needed	
Demo	• Old plastic doll with hair, such as a Barbie or pony doll, Tub of water
Projects	• Egg carton, Construction paper, Construction paper, Paper clips, Magnet, Dowel rod, String

Chapter 16: A Dip in the Ocean

Read: Gathering Information

Living Book Reading Assignment

📖 Chapter 16 of *The Sassafras Science Adventures Volume 1: Zoology*

(Optional) Encyclopedia Readings

🔖 *Kingfisher First Encyclopedia of Animals* p. 92 (Penguin), p. 118 (Fish)

🔖 *DK Encyclopedia of Animals* pp. 276-277 (Penguin), pp. 174-176 (Fish), pp. 74-75 (Ocean)

(Optional) Additional Library Books

📖 *Face to Face with Penguins* (Face to Face with Animals) by Yva Momatiuk

📖 *Emperor Penguin* (Life Cycle of A...) by Bobbie Kalman

📖 *National Geographic Readers: Penguins!* by Anne Schreiber

📖 *What's It Like to Be a Fish?* (Let's-Read-and-Find... Science 1) by Wendy Pfeffer

📖 *Where Fish Go In Winter* by Amy Goldman Koss and Laura J. Bryant

📖 *The Life Cycle of Fish* (Life Cycles) by Darlene R. Stille

📖 *Eye Wonder: Ocean* by Sue Thornton and Mary Ling

📖 *Over in the Ocean: In a Coral Reef* (Sharing Nature With Children) by Marianne Berkes and Jeanette Canyon

📖 *A Swim Through the Sea* by Kristin Joy Pratt-Serafini

Write: Keeping a Notebook

SCIDAT Logbook Sheets

This week, you can have the students fill out a logbook page for the penguin and the codfish, as well as begin to fill out a Habitat Information Sheet for the Ocean. The students could use the following information:

Habitat Information Sheet - Ocean

HABITAT: Ocean

LOCATION AND LOCAL EXPERT: Atlantic Ocean and James Q. McScruffy

AVERAGE RAINFALL

AVERAGE TEMPERATURE: It ranges from cool to cold.

MAIN CHARACTERISTICS: The ocean has lots of water.

ANIMALS FOUND THERE: Add the penguin and fish.

Animal Record Sheet - Penguin

ANIMAL NAME: King Penguins

CLASSIFICATION: Bird

FOOD: Carnivore

LOCATION FOUND: South Georgia Island (or Atlantic Ocean)

INFORMATION LEARNED

- They are the second largest species of penguin, found on the South Georgia Island and along the coast of Antarctica.
- They have golden-orange patches on their ears, bills and upper breasts.
- They are birds that can't fly.
- They are excellent swimmers and seem to fly through the ocean.
- They use their stiff wings as flippers and their tails and feet for steering.
- Penguins have waterproof feathers and a thick layer of fat to help them stay dry and warm in the icy waters.
- They can jump above the surface of the water, like a dolphin, so that they can breathe when swimming quickly or to get back up on land after being in the water.
- They hunt and eat fish and krill.
- They have spiky tongues that help them to grip slippery fish.
- Penguins are black and white as a means of camouflage.
- When they are swimming, they appear dark from the top and pale from below, making it harder for predators to spot from either vantage point.
- They come out of the water to lay their eggs and raise their chicks.
- They lay one to two eggs at a time and both the mother and father will take turns incubating the egg.
- Once they hatch, the chicks will huddle together while their parents go to the sea for food.
- The chicks cannot go into the ocean until they have grown their waterproof adult feathers.

Animal Record Sheet - Fish

ANIMAL NAME: Codfish

CLASSIFICATION: Fish

FOOD: Carnivore

LOCATION FOUND: Atlantic Ocean

INFORMATION LEARNED

- Codfish have a backbone and a basic skeleton.
- They don't have a swim bladder to control their buoyancy instead, they have a reduced amount of minerals in their bones and increased fatty tissue that gives them a near neutral density.
- They are covered with scales, which help to reduce drag as they swim through the water.
- They have strong fins that they use to move through the water.
- They need oxygen to breathe, but instead of breathing in air, they absorb oxygen from the water as it passes over their gills.
- They have a chemical in their blood that acts like antifreeze, plus their spleen is able to filter ice crystals from their blood that might form.

- They are cold-blooded, which means that they cannot regulate their own body temperature.
- They eat small fish and crustaceans, such as krill, which makes them carnivores.
- They often swim in groups called schools; they mainly do this for safety because many fish together can confuse a predator and make it hard for them to single out and catch one fish.

VOCABULARY

Have your older students look up the following terms in the glossary in the appendix on pp. 145-146 or in a science encyclopedia. Have them copy the definitions onto a blank index card or into their SCIDAT logbook.

- ✍ OCEAN – A habitat characterized by water as the basis and support for all life; plants are typically a species of algae.
- ✍ FISH – A cold-blooded, aquatic vertebrate that has gills, fins, and typically an elongated body covered with scales.

(OPTIONAL) COPYWORK

Copywork Sentence

Fish are cold-blooded animals that are covered with scales.

Dictation Selection

The ocean is a unique salt-water habitat that covers about two-thirds of the Earth. It supports different kinds of marine animals that have adapted to life under the water. The main species of plants found in the ocean are algae.

DO: PLAYING WITH SCIENCE

SCIENTIFIC DEMONSTRATION: FISH GILLS

Materials
- ☑ Old plastic doll with hair, such as Barbie or pony doll (NOTE—If you don't have any dolls with hair, you can use an old magazine instead.)
- ☑ Tub of water

Procedure
1. Place your doll's hair under the water. Observe how the hair flows and then try to separate out several strands of hair.
2. Pull the doll out of the water. Once again, observe how the hair flows and then try to separate out several strands of hair.

Explanation

The students should see that when the hair was under the water it flowed freely and it was very easy to separate several strands. Once the hair was out of the water it stuck together in clumps and it took some effort to be able to separate out several strands. This is the same with fish gills. Under water they can easily float and move, making it easy for the fish to move water through their gills and absorb the oxygen they need. When the fish is out of water, their gills clump together and it is very hard for them to move anything past their gills.

Take It Further

Visit an aquarium with the students, or set up your own so the students can observe real fish gills in action!

(OPTIONAL) STEAM PROJECTS

Multi-chapter Activities

✂ FOOD CHART – This week, add the penguin and the fish to the carnivore side of your food chart. You can use the mini-animal pictures found in the appendix on p. 129 of this guide or print out your own.

✂ HABITAT PROJECT – Make a poster or diorama that depicts the ocean. This week, add the penguin and the fish. You can use the mini-animal pictures found in the appendix on p. 129 of this guide and the habitat poster on p. 137 or print out your own.

Activities For This Chapter

✂ MAKE A PENGUIN – You will need several egg carton cups, construction paper and paint. Use the directions from the following website:

🖰 http://www.dltk-kids.com/animals/mcarton-penguin.htm

✂ GO FISHING – Begin by cutting out 5 red, 4 blue, 3 green and 2 yellow fish. Attach a paper clip to the end of each of the fish and place them all into a bucket or box that is not transparent. Next, tie a magnet to a string and tie the string to a pole, either a dowel rod or a cardboard tube. Then, go fishing and see what you catch. The red fish are worth 1 point each, the blue fish are worth 2 points each, the green fish are worth 5 points each and the yellow fish are worth 10 points each. The person with the most points wins the game.

CHAPTER 17: GRID SCHEDULE

Supplies Needed	
Demo	• Shallow bowl, Water, Digital camera
Projects	• 2 Gallon plastic jug, Googly eyes, Hot dogs

Chapter Summary

The chapter opens with Blaine and Tracey discussing some of the information they have learned from Captain James Q. McScruffy. Then, they spot something shimmering in the water, something that is mistaken for treasure, but in reality it is hundreds of stinging jellyfish. They continue to spot signs from the poem that the captain's great-grandfather left, along with a blue whale that the captain shares more about. As they continue on their way, William is able to convince the captain to share more about squids for the twins. After that, they pass the Shivering Timbers Island, and soon after they spot the floating treasure. Captain James Q. McScruffy goes after it and ends up fighting the giant squid, but once again it disappears along with the treasure chest. The chapter ends with the Sassafras twins thinking that their adventure is coming to a close because their LINLOC app says that they are heading back to Uncle Cecil's lab.

Weekly Schedule

	Day 1	Day 2	Day 3	Day 4
Read	☐ Read the section entitled "Watching Whales" of Chapter 17 in *SSA Volume 1: Zoology.*	☐ Read the section entitled "The Shivering Squid" of Chapter 17 in *SSA Volume 1: Zoology.*	☐ (*Optional*) Read one or all of the assigned pages from the encyclopedia of your choice.	☐ (*Optional*) Read one of the additional library books.
Write	☐ Fill out the Animal Record Sheet on SL p. 81 for the whale. ☐ Go over the vocabulary words and enter them into the Zoology Glossary on SL p. 98.	☐ Fill out the Animal Record Sheet on SL p. 82 for the squid. ☐ Add to the Habitat Information Sheet on SL p. 77 for the Ocean.	☐ Write the information learned from the demonstration on SL p. 84. ☐ (*Optional*) Write a narration on the Zoology Notes Sheet on SL p. 84.	☐ (*Optional*) Complete the copywork or dictation assignment and add it to the Zoology Notes sheet on SL p. 84. ☐ (*Optional*) Take Zoology Quiz #8.
Do	☐ (*Optional*) Make a Whale Water Scooper.	☐ (*Optional*) Make Squid Dogs.	☐ Do the demonstration entitled "Echolocation."	☐ (*Optional*) Add the animals studied this week to the food chart and habitat posters.

CHAPTER 17: LIST SCHEDULE

CHAPTER SUMMARY

The chapter opens with Blaine and Tracey discussing some of the information they have learned from Captain James Q. McScruffy. Then, they spot something shimmering in the water, something that is mistaken for treasure, but in reality it is hundreds of stinging jellyfish. They continue to spot signs from the poem that the captain's great-grandfather left, along with a blue whale that the captain shares more about. As they continue on their way, William is able to convince the captain to share more about squids for the twins. After that, they pass the Shivering Timbers Island, and soon after they spot the floating treasure. Captain James Q. McScruffy goes after it and ends up fighting the giant squid, but once again it disappears along with the treasure chest. The chapter ends with the Sassafras twins thinking that their adventure is coming to a close because their LINLOC app says that they are heading back to Uncle Cecil's lab.

ESSENTIALS

Read

- ☐ Read the section entitled "Watching Whales" of Chapter 17 in *SSA Volume 1: Zoology*.
- ☐ Read the section entitled "The Shivering Squid" of Chapter 17 in *SSA Volume 1: Zoology*.

Write

- ☐ Fill out the Animal Record Sheet on SL p. 81 for the whale.
- ☐ Go over the vocabulary words and enter them into the Zoology Glossary on SL p. 98.
- ☐ Fill out the Animal Record Sheet on SL p. 82 for the squid.
- ☐ Add to the Habitat Information Sheet on SL p. 77 for the Ocean.
- ☐ Write the information learned from the demonstration on SL p. 84.

Do

- ☐ Do the demonstration entitled "Echolocation."

(OPTIONAL) EXTRAS

Read

- ☐ Read one of the additional library books.
- ☐ Read one or all of the assigned pages from the encyclopedia of your choice.

Write

- ☐ Write a narration on the Zoology Notes Sheet on SL p. 84.
- ☐ Complete the copywork or dictation assignment and add it to the Zoology Notes sheet on SL p. 84.
- ☐ Take Zoology Quiz #8.

Do

- ☐ Make a Whale Water Scooper.
- ☐ Make Squid Dogs.
- ☐ Add the animals studied this week to the food chart and habitat posters.

Supplies Needed	
Demo	• Shallow bowl, Water, Digital camera
Projects	• 2 Gallon plastic jug, Googly eyes, Hot dogs

Chapter 17: Arr, Treasure Ahead!

Read: Gathering Information

Living Book Reading Assignment

📖 Chapter 17 of *The Sassafras Science Adventures Volume 1: Zoology*

(Optional) Encyclopedia Readings

🔖 *Kingfisher First Encyclopedia of Animals* p. 70 (Whale), p. 129 (Octopus and Squid)

🔖 *DK Encyclopedia of Animals* pp. 353-354 (Whale), pp. 255-257 (Octopus and Squid)

(Optional) Additional Library Books

📖 *Face to Face with Whales* (Face to Face with Animals) by Flip Nicklin

📖 *Amazing Whales!* (I Can Read Book 2) by Sarah L. Thomson

📖 *Is a Blue Whale the Biggest Thing There Is?* (Robert E. Wells Science) by Robert E. Wells

📖 *Squids* (Blastoff! Readers: Oceans Alive) by Colleen A. Sexton

📖 *Squids* (Under the Sea) by Rake and Jody S.

📖 *Squirting Squids* (No Backbone! The World of Invertebrates) by Natalie Lunis

Write: Keeping a Notebook

SCIDAT Logbook Sheets

This week, you can have the students fill out a logbook page for the whale and the squid, as well as continue to fill out a Habitat Information Sheet and the Around the World Sheet for the Ocean. The students could use the following information:

Habitat Information Sheet - Ocean

HABITAT: Ocean

LOCATION AND LOCAL EXPERT

AVERAGE RAINFALL

AVERAGE TEMPERATURE

MAIN CHARACTERISTICS: The ocean covers two-thirds of the Earth. It is made up of salt water.

ANIMALS FOUND THERE: Add the whale and squid (The students could also add jellyfish).

Around the World - World Oceans

MAP: Have the students put a star on the South Georgia Island in the Atlantic Ocean and label it "South Georgia Island". Then, have them color all of the oceans blue and label

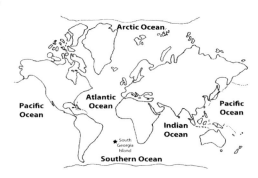

each of them (i.e. Pacific Ocean, Atlantic Ocean, Indian Ocean, and Southern Ocean). Use the map pictured as a guide.

Animal Record Sheet - Whale

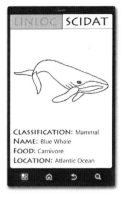

ANIMAL NAME: Blue Whale

CLASSIFICATION: Mammal

FOOD: Carnivore

LOCATION FOUND: Atlantic Ocean

INFORMATION LEARNED

- Whales are mammals, not fish.
- They also feed their young with milk.
- They are warm-blooded and breathe air through the lungs.
- They are almost hairless, but they do have a thick layer of blubber to keep them warm.
- Whales are the largest animals in the world, with the blue whale topping out at over one hundred feet long and weighing one hundred tons-the size of fifteen elephants.
- Their massive weight is supported by the water and if they are stranded on land, their weight will crush their internal organs.
- They belong to a group of whales known as baleen whales, which means they catch their food by filtering the water around them through a comb like plate in their mouth that is equipped with bristles that capture the food.
- They feed on the tiniest creature in the ocean, krill.
- The blue whale can eat up to two and a half tons of krill each day.
- They move through the water using their tail fins.
- Every hour or so, the whale must surface to new air.
- They blow stale air and water out of their two blowholes on the top of their head and take in fresh air.
- Their spout can go nearly thirty feet in the air.
- The blue whale will migrate towards the equator as winter approaches.
- The average lifespan of a blue whale is eighty to ninety years.

Animal Record Sheet - Squid

ANIMAL NAME: Squid

CLASSIFICATION: Invertebrate

FOOD: Carnivore

LOCATION FOUND: Atlantic Ocean

INFORMATION LEARNED

- They are related to the octopus, except they have ten arms instead of eight.
- Two of their arms are a bit longer and have suckers all along them, which they use for capturing prey; it eats mostly crabs, fish, and shellfish.
- There are all different types of squid that range in length from less than an inch to up to sixty-five feet.
- They get around using jet propulsion.
- They squirt water through their siphon, which is a muscular tube behind the head that can swivel to help them go in the direction they want.

- They can change color to match their surroundings or to show emotion.
- They have an ink gland.
- When they are threatened, they squirt out a cloud of ink into the water that confuses their attackers.
- Squids are found at every level of the ocean.
- They have keen eyesight, large brains and fast reaction times.
- They are also some of the fastest swimmers in the water.

VOCABULARY

Have your older students look up the following term in the glossary in the appendix on pp. 145-146 or in a science encyclopedia. Have them copy the definition onto a blank index card or into their SCIDAT logbook.

✐ INVERTEBRATE – An animal without a vertebral column.

(OPTIONAL) COPYWORK

Copywork Sentence

An invertebrate is an animal that does not have a backbone.

Dictation Selection

Invertebrates are animals that do not have a vertebral column, or backbone. They include animals such as worms, insects, spiders, clams, and squid. The majority of animals on earth are invertebrates.

(OPTIONAL) QUIZ

This week, you can give the students a quiz based on what they learned in chapters 16 and 17. You can find the quiz in the appendix on p. 169.

Quiz #8 Answers
1. C
2. D
3. A
4. B
5. D

DO: PLAYING WITH SCIENCE

SCIENTIFIC DEMONSTRATION: ECHOLOCATION

Materials
- ☑ Shallow bowl
- ☑ Water
- ☑ Digital camera

Procedure
1. Fill a shallow bowl with water.
2. Have the student touch the center of the water with their fingertip as you take a picture.
3. Look at the picture and ask the students:

? What do you see?

Explanation

The students should see circular waves radiate out from the point in the water that they touched. Sounds waves travel in the same manner through the water. Whales use this principle of sound waves to communicate with one another, to locate their pod, and to hunt for food. This process of using sound to locate an object is known as echolocation.

Take It Further

Have the students try to communicate with each other using only clicks and whistles.

(OPTIONAL) STEAM PROJECTS

Multi-chapter Activities

✂ FOOD CHART – This week, add the whale and the squid to the carnivore side of your food chart. You can use the mini-animal pictures found in the appendix on p. 129 of this guide or print out your own.

✂ HABITAT PROJECT – This week, add the whale and the squid to your ocean habitat. You can use the mini-pictures found in the appendix on p. 129 of this guide or print out your own.

Activities For This Chapter

✂ WHALE WATER SCOOPER – Make a whale water scooper using a rinsed out 2 gallon jug. Begin by mixing 4 parts grey paint and 1 part glue and use the mixture to paint the jug all over. Then, cut out a curved line in the bottom of the jug. Next, glue on two googly eyes just above the curved line on either side of the jug and cut a small blow hole between the eyes and the handle. Finally, cut a long skinny heart out of a sheet of grey foam and attach it to the top of the jug for the whale's tail.

✂ SQUID DOGS – Make squid-shaped hot dogs for lunch. Begin by cutting a hot dog two-thirds of the way up several times to create legs. Then, cut the top third of the hot dog so that it looks like a triangle. Next, cook the hot dog on the stove according to the package instructions. Serve with the students' favorite side!

CHAPTER 18: GRID SCHEDULE

Supplies Needed	
Demo	• Zoology Bingo Cards (Download for free from Elemental Science)
Projects	• Blank map

Chapter Summary

The chapter opens up with the Man With No Eyebrows recalling how he has been unsuccessful in stopping the twins, but he vows to keep trying. It then flashes back to Blaine and Tracy arriving back at Uncle Cecil's basement laboratory. They share some of their journey with their uncle and they receive bonus data on animal defenses, diet and migration. Tracey sees a ProLog hat on Uncle Cecil's desk and after some discussion they all realize that the Man With No Eyebrows must have access to his own invisible zip lines. The chapter ends with Uncle Cecil sharing with the twins that their adventure is only just beginning.

Weekly Schedule

	Day 1	**Day 2**	**Day 3**	**Day 4**
Read	☐ Read "Bonus Data" of Chapter 18 in *SSA Volume 1: Zoology*.	☐ Read the section entitled "A Fun Surprise" of Chapter 18 in *SSA Volume 1: Zoology*.	☐ (*Optional*) Read one or all of the assigned pages from the encyclopedia of your choice or read one of the library books.	☐ Review the work you have done over the unit.
Write	☐ Add to the Bonus Data Sheet on SL p. 87.	☐ Go over the vocabulary word and enter it into the Zoology Glossary on SL p. 99. ☐ (*Optional*) Write a narration on the Zoology Notes Sheet on SL p. 87.	☐ (*Optional*) Write a narration on the Zoology Notes Sheet on SL p. 88.	
Do	☐ (*Optional*) Do the Migration Routes project.		☐ Play a game of Zoology Bingo.	

CHAPTER 18: LIST SCHEDULE

CHAPTER SUMMARY

The chapter opens up with the Man With No Eyebrows recalling how he has been unsuccessful in stopping the twins, but he vows to keep trying. It then flashes back to Blaine and Tracy arriving back at Uncle Cecil's basement laboratory. They share some of their journey with their uncle and they receive bonus data on animal defenses, diet and migration. Tracey sees a ProLog hat on Uncle Cecil's desk and after some discussion they all realize that the Man With No Eyebrows must have access to his own invisible zip lines. The chapter ends with Uncle Cecil sharing with the twins that their adventure is only just beginning.

ESSENTIALS

Read

- ☐ Read "Bonus Data" of Chapter 18 in *SSA Volume 1: Zoology.*
- ☐ Read the section entitled "A Fun Surprise" of Chapter 18 in *SSA Volume 1: Zoology.*

Write

- ☐ Add to the Bonus Data Sheet on SL p. 87.
- ☐ Go over the vocabulary word and enter it into the Zoology Glossary on SL p. 99.

Do

- ☐ Play a game of Zoology Bingo.
- ☐ Review the work you have done over the unit.

(OPTIONAL) EXTRAS

Read

- ☐ Read one of the additional library books.
- ☐ Read one or all of the assigned pages from the encyclopedia of your choice.

Write

- ☐ Write a narration on the Zoology Notes Sheet on SL p. 87.
- ☐ Complete the copywork or dictation assignment and add it to the Zoology Notes sheet on SL p. 88.

Do

- ☐ Do the Migration Routes project.

Supplies Needed	
Demo	• Zoology Bingo Cards (Download for free from Elemental Science)
Projects	• Blank map

Chapter 18: The Adventure Ends... Or Does it?

READ: Gathering Information

Living Book Reading Assignment

📖 Chapter 18 of *The Sassafras Science Adventures Volume 1: Zoology*

(Optional) Encyclopedia Readings

ያ *Kingfisher First Encyclopedia of Animals* pp. 10-11 (Defense), p. 12 (Food), p. 9 (Migration)
ያ *DK Encyclopedia of Animals* pp. 34-37 (Defense), pp. 30-33 (Food), pp. 78-79 (Migration)

(Optional) Additional Library Books

📖 *Discover Science: Animal Disguises* by Belinda Weber
📖 *Fake Out!: Animals That Play Tricks* (All Aboard Science Reader) by Ginjer L. Clarke and Pete Mueller
📖 *What Are Food Chains and Webs?* (Science of Living Things) by Bobbie Kalman and Jacqueline Langille
📖 *Who Eats What? Food Chains and Food Webs* (Let's-Read-and-Find... Science, Stage 2) by Patricia Lauber and Holly Keller
📖 *DK Readers: The Great Migration* by Deborah Lock
📖 *Going Home: The Mystery of Animal Migration* (Sharing Nature with Children Books) by Marianne Berkes and Jennifer DiRubbio

WRITE: Keeping a Notebook

SCIDAT Logbook Sheets

This week, you can have the students narrate what they have learned about methods of defense and migration on the bonus data sheets. The students could use the following information:

Bonus Data - Methods of Defense

- **PLAYING DEAD:** The prey acts as if they are already dead, which can shut off the hunting behavior in a predator.
- **MAKING AN ESCAPE:** The prey makes a sudden dash for safety. This method relies on the fact that the prey is fast and has sharp senses.
- **SPINES OR SCALES:** The prey is covered with spines or scales that make it difficult for the predator to eat it.
- **CAMOUFLAGE:** The prey disguises itself as something else so that the predator cannot find it.
- **MIMICRY:** The prey mimics another more dangerous animal so that the predator will leave it alone.
- **CHEMICAL WEAPONS:** The prey emits a poisonous or foul-smelling chemical to keep predators from eating it.

Bonus Data - Diet

- Carnivores are meat-eaters.
- Herbivores are plant-eaters.
- Omnivores are both meat- and plant-eaters.
- Food chains are chains of living things that eat each other.
- A food web is a network of food chains for a given ecosystem. Autotrophs, such as plants, make their own food by using energy from the sun.
- Heterotrophs, such as fungi and animals, get their energy from the food they eat.
- Autotrophs and heterotrophs are dependent on each other.

Bonus Data - Migration

- Animals migrate over land, water, and air, making annual journeys to find better living conditions.
- Some travel short distances; others travel from one end of the globe to the other.
- Birds migrate using an internal compass that relies on the sun and stars to help with position.
- Other animals find their way by following their parents or the rest of the herd.

VOCABULARY

Have your older students look up the following term in the glossary in the appendix on pp. 145-146 or in a science encyclopedia. Have them copy the definition onto a blank index card or into their SCIDAT logbook.

↻ MIGRATION — A journey by an animal to a new habitat.

(OPTIONAL) COPYWORK

Copywork Sentence

Animals migrate over land, water, and air.

Dictation Selection

Animals migrate over land, water, and air to find better living conditions. Some animals travel short distances, like the reindeer. Other animals travel from one end of the globe to the other, like the arctic tern. Birds migrate using an internal compass that relies on the sun and stars to help with position.

DO: PLAYING WITH SCIENCE

REVIEW GAME: ZOOLOGY BINGO

Materials
- ☑ Zoology Bingo from Elemental Science

Procedure
1. Download the game templates from the following website:
   ↻ https://elementalscience.com/collections/free-printable-games/products/zoology-bingo
2. Play Zoology Bingo according to the directions included in the game packet.

Activities For This Chapter

✄ MIGRATION ROUTES – After you read about migration, use the blank map in the student logbook to trace the migration routes of the swallow, reindeer, monarch butterfly, arctic tern, and humpback whale. Use a different color or style of line for each of the animals. The map pictured below is a guide.

Migration Routes

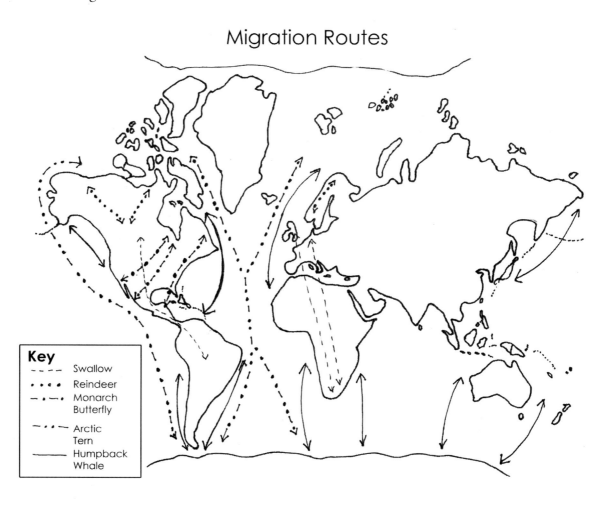

Key
- – – – Swallow
- • • • • Reindeer
- — • — • Monarch Butterfly
- — • • — Arctic Tern
- ———— Humpback Whale

APPENDIX

LAB REPORT SHEET

Title

_____

Hypothesis (What I Think Will Happen)

_____

_____

_____

Materials (What We Used)

_____   _____

_____   _____

_____   _____

Procedure (What We Did)

_____

_____

_____

_____

_____

_____

_____

_____

_____

_____

_____

Observations and Results (What I Saw and Measured)

_____

_____

_____

_____

_____

_____

_____

_____

_____

_____

Conclusion (What I Learned)

_____

_____

_____

_____

_____

_____

_____

Small Animal Pictures

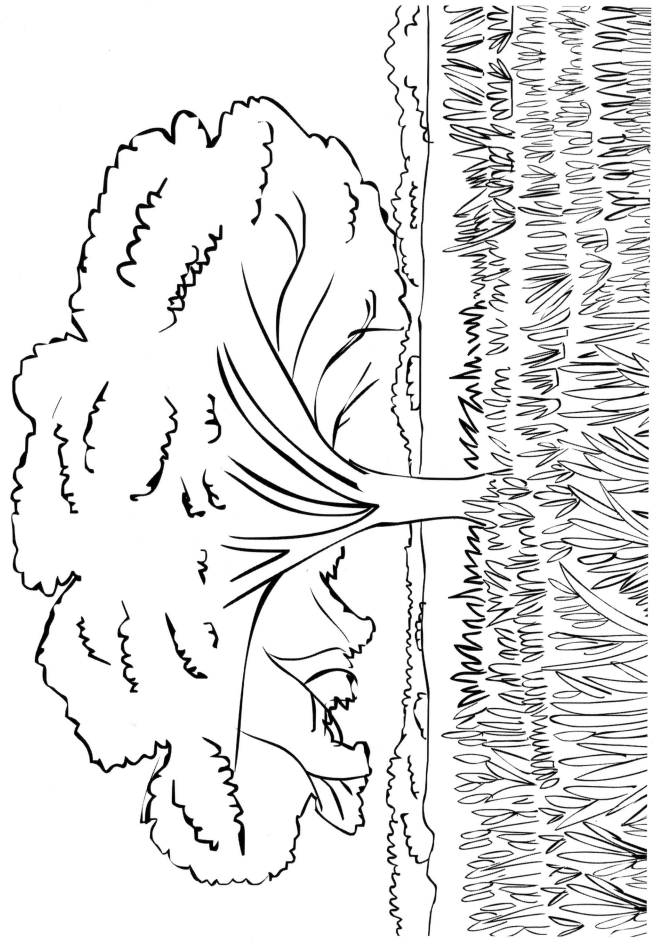

The Sassafras Guide to Zoology ~ Appendix

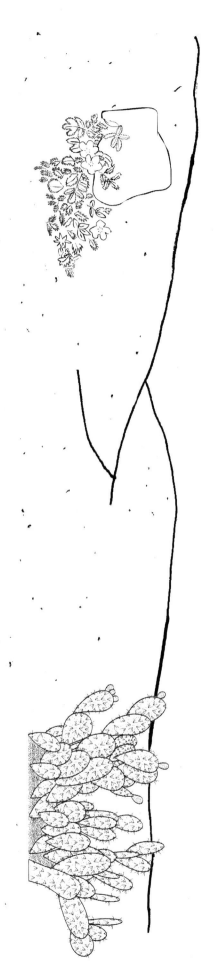

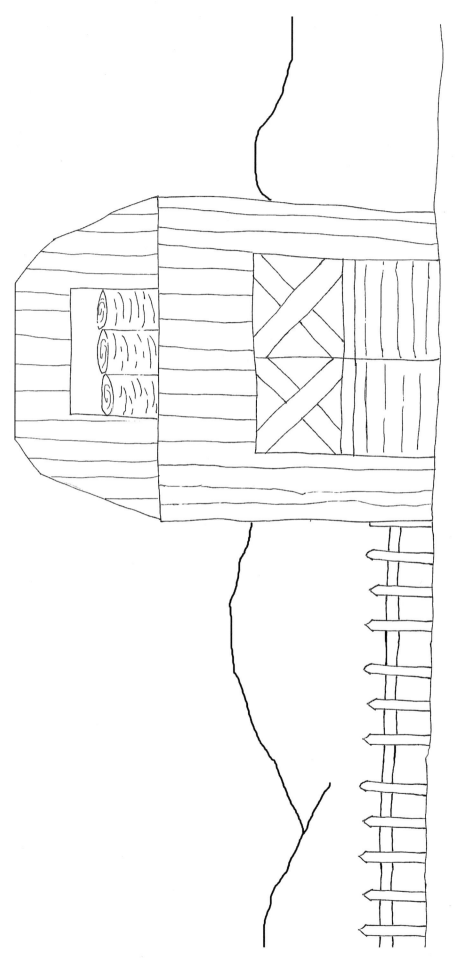

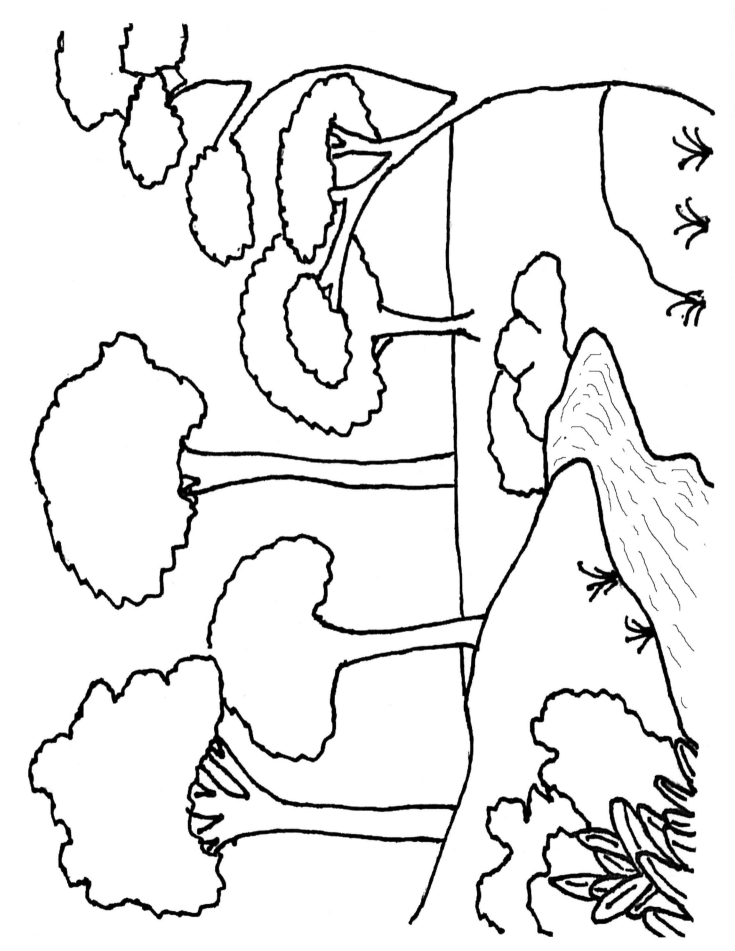

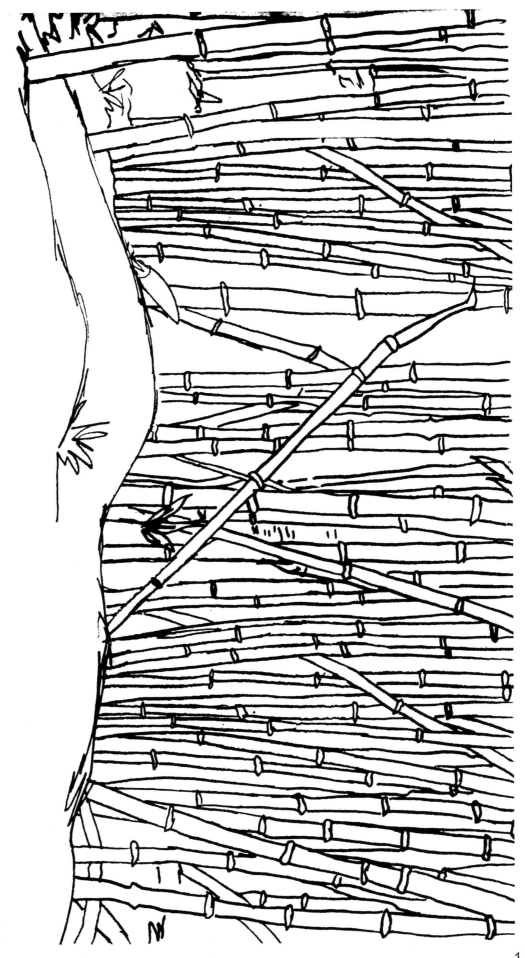

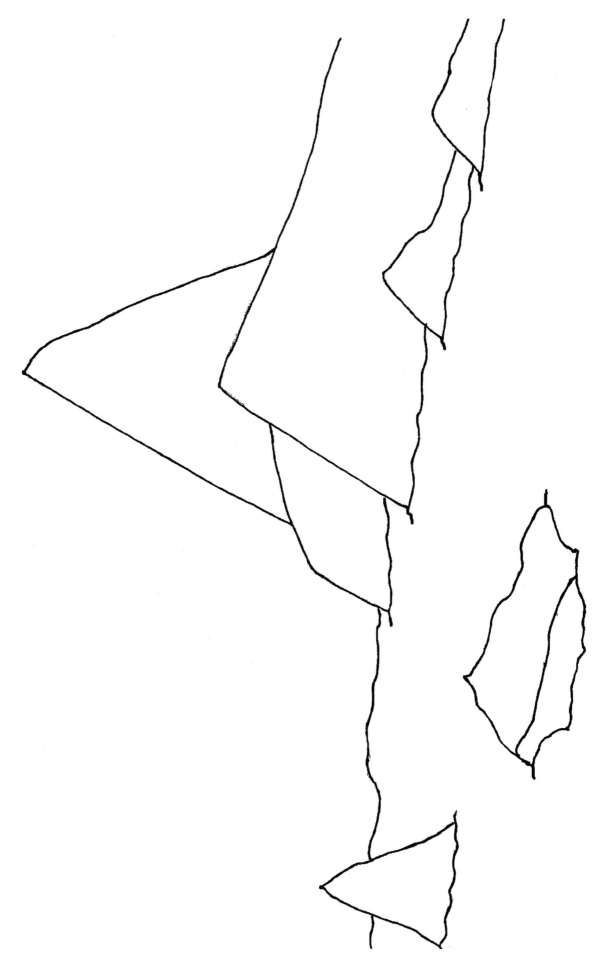

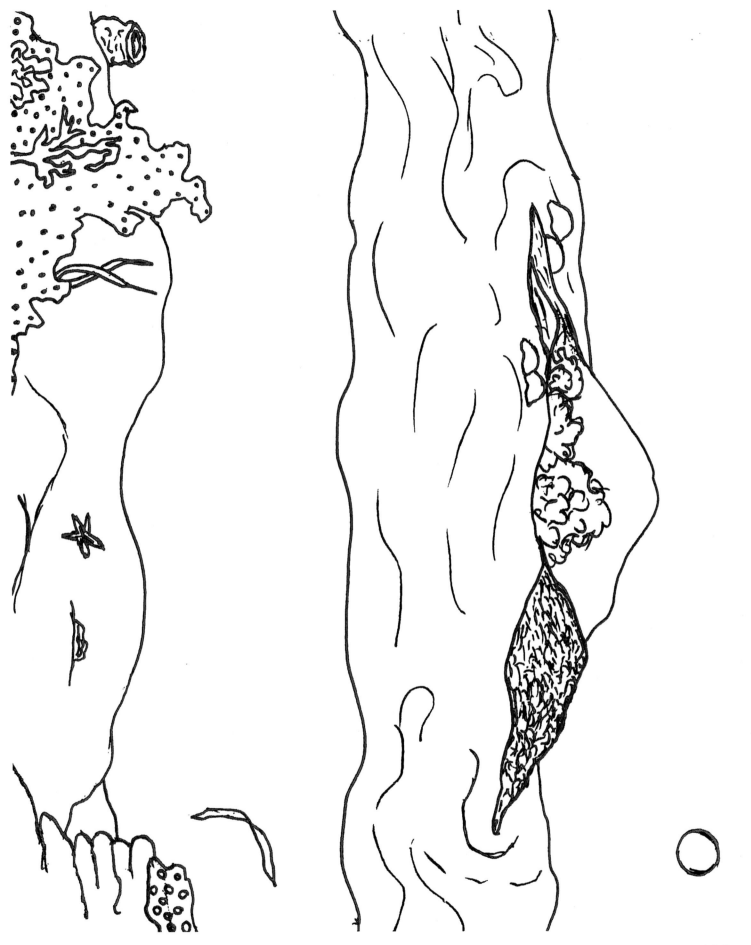

TOUCAN

Life Cycle of a Frog

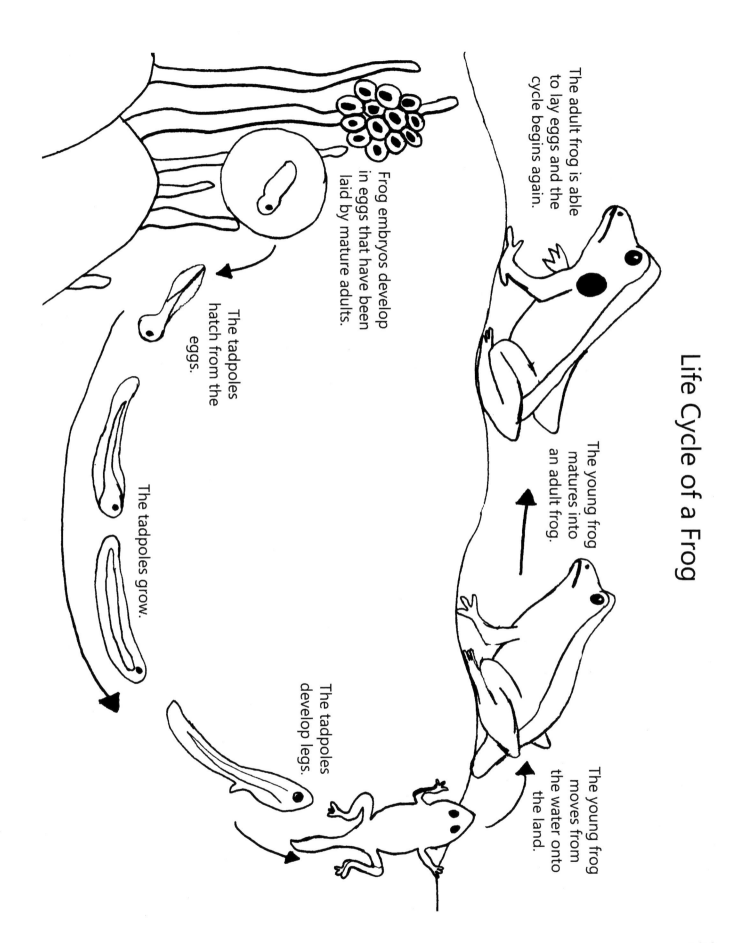

Frog embryos develop in eggs that have been laid by mature adults.

The tadpoles hatch from the eggs.

The tadpoles grow.

The tadpoles develop legs.

The young frog moves from the water onto the land.

The young frog matures into an adult frog.

The adult frog is able to lay eggs and the cycle begins again.

Life Cycle of a Frog

Life Cycle of a Butterfly

Butterflies lay eggs on leaves.

Caterpillars hatch out of the eggs and eat the leaves.

When they are full, caterpillars make a chrysalis.

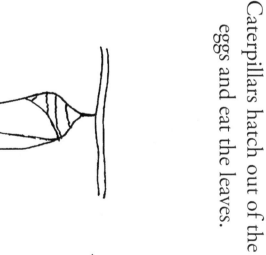

A butterfly emerges from the chrysalis.

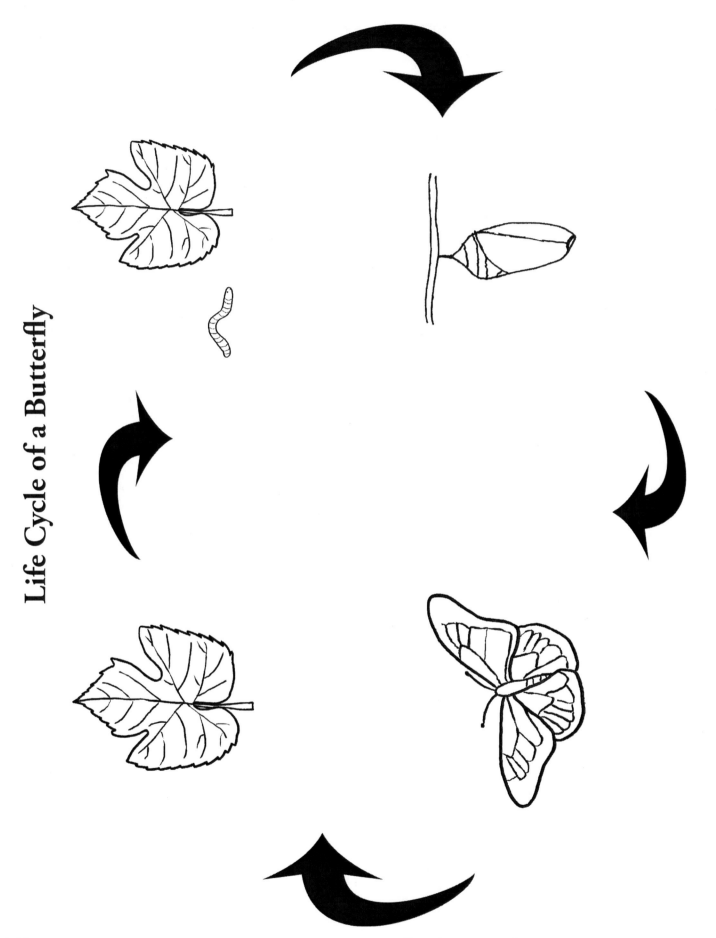

Life Cycle of a Butterfly

GLOSSARY

Zoology Glossary

A

- AMPHIBIAN – A cold-blooded, smooth-skinned vertebrate, such as a frog or salamander.

- ARCTIC – A habitat that has little vegetation and very cold temperatures.

- ARTHROPOD – An animal with a jointed body, such as an insect or spider.

B

- BIRD – A warm-blooded, egg-laying, feathered vertebrate; it also has wings.

C

- CARNIVORE – An animal that eats meat.

- CLASSIFICATION – A way of identifying or grouping living things.

D

- DESERT – A habitat that receives very little rain; it is hot during the day and cold at night.

- DOMESTICATED ANIMAL – An animal that has been under human control for many generations.

E

F

- FISH – A cold-blooded, aquatic vertebrate that has gills, fins, and typically an elongated body covered with scales.

- FOOD CHAIN – A chain of living things that eat each other.

- FOREST – A habitat that is characterized by the abundance of trees, either deciduous or evergreen.

G

- GRASSLAND – A habitat characterized by vast grassy fields.

H

- HERBIVORE – An animal that eats plants.

- HIBERNATION – The ability of an animal to go into a deep sleep for a long period of time by shutting down its body.

I

- INSECT – An animal that has three body parts (head, thorax, and abdomen) and six legs.

- INVERTEBRATE – An animal without a backbone.

J

K

L

M

- MAMMALS – Any warm-blooded vertebrate with skin that is more or less covered with hair; they give birth to live young which are nourished with milk at the beginning of their life.

- MARSUPIAL – A group of mammals that give birth to immature young that complete their development in the mother's pouch.

- MIGRATION – A journey by an animal to a new habitat.

N

O

- OBSERVATION – Something that you see with your eyes, or the act of regarding attentively.

- OCEAN – A habitat characterized by water as the basis and support for all life; plants are typically a species of algae.

- OMNIVORE – An animal that eats both plants and meat.

P

Q

R

- RAINFOREST – A habitat with lots of plants, trees, and animals due to the heavy amount of rain it receives.

- REPTILE – A group of cold-blooded animals that usually have rough skin.

S

T

U

V

- VERTEBRATE – An animal that has an internal skeleton made from bone.

W

X

Y

Z

BOOK LIST

ADDITIONAL LIBRARY BOOKS
LISTED BY CHAPTER

CHAPTER 1

  📖 *What Is the Animal Kingdom?* (Science of Living Things) by Bobbie Kalman

  📖 *Animal Classification* by Polly Goodman

  📖 *Who's in Your Class?* Level 4: An Animal Adventure (Lithgow Palooza Readers: Level 4) by John Lithgow and Susan Blackaby

CHAPTER 2

  📖 *Face to Face with Lions* (Face to Face with Animals) by Dereck Joubert and Beverly Joubert

  📖 *Tawny Scrawny Lion* (Little Golden Book) by Golden Books and Gustaf Tenggren

  📖 *The Cheetah: Fast as Lightning* (Animal Close-Ups) by Christine Denis-Huot and Michel Denis-Huot

  📖 *Cheetah* (Welcome Books: Animals of the World) by Edana Eckart

  📖 *Cheetah Cubs: Station Stop 2* (All Aboard Science Reader) by Ginjer L. Clarke and Lucia Washburn

  📖 *What is a Mammal?* (Science of Living Things) by Kalman

CHAPTER 3

  📖 *Elephants: A Book for Children* by Steve Bloom

  📖 *Face to Face With Elephants* (Face to Face with Animals) by Beverly Joubert

  📖 *Giraffes* by Jill Anderson

  📖 *Baby Giraffes* (It's Fun to Learn about Baby Animals) by Bobbie Kalman

  📖 *Chee-Lin: A Giraffe's Journey* by James Rumford

CHAPTER 4

  📖 *Camels* (Nature Watch) by Cherie Winner

  📖 *I Wonder Why Camels Have Humps: And Other Questions About Animals* by Anita Ganeri

  📖 *King Cobras* by Joanne Mattern and Gail Saunders-Smith

  📖 *Egyptian Cobra* (Killer Snakes) by Jessica O'Donnell

  📖 *Smart Kids Reptiles* by Simon Mugford

CHAPTER 5

  📖 *Fun Facts About Lizards!* (I Like Reptiles and Amphibians!) by Carmen Bredeson

  📖 *All About Lizards* by Jim Arnosky

  📖 *Fox* by Kate Banks

  📖 *Foxes* (Animal Predators) by Sandra Markle

  📖 *Desert Animals* (Animals in Their Habitats) by Francine Galko

  📖 *Life in the Desert* (Pebble Plus: Habitats Around the World) by Alison Auch

  📖 *Baby Animals in Desert Habitats* (Habitats of Baby Animals) by Bobbie Kalman

CHAPTER 6

  📖 *Cows and Their Calves* (Pebble Plus: Animal Offspring) by Margaret Hall

  📖 *Raising Cows on the Koebels' Farm* (Our Neighborhood) by Alice K. Flanagan

  📖 *Milk: From Cow to Carton* (Let's-Read-and-Find... Book) by Aliki

  📖 *Are You a Bee?* (Backyard Books) by Judy Allen

  📖 *The Life and Times of the Honeybee* by Charles Micucci

- 📖 *DK Readers: Busy, Buzzy Bee* (Level 1: Beginning to Read) by Karen Wallace
- 📖 *What if There Were No Bees?* by Suzanne Buckingham Slade and Carol Schwartz

CHAPTER 7

- 📖 *From Egg to Chicken* (How Living Things Grow) by Anita Ganeri
- 📖 *Chickens Aren't the Only Ones* (World of Nature Series) by Ruth Heller
- 📖 *Chickens* (Animals That Live on the Farm) by JoAnn Early Macken
- 📖 *Time For Kids: Spiders!* by Editors of TIME For Kids
- 📖 *Spinning Spiders* (Let's-Read-and-Find... Science 2) by Melvin Berger
- 📖 *The Very Busy Spider* by Eric Carle

CHAPTER 8

- 📖 *"Slowly, Slowly, Slowly," said the Sloth* by Eric Carle
- 📖 *Baby Sloth* (Nature Babies) by Aubrey Lang and Wayne Lynch
- 📖 *Sloths* (Animals That Live in the Rain Forest) by Julie Guidone
- 📖 *Score One for the Sloths* by Helen Lester
- 📖 *Toucans and Other Birds* (Animals That Live in the Rain Forest) by Julie Guidone
- 📖 *Toucans* (Pebble Plus) by Mary R. Dunn
- 📖 *Toco Toucans: Bright Enough to Disappear* (Disappearing Acts) by Anastasia Suen
- 📖 *A Rainforest Habitat* (Introducing Habitats) by Molly Aloian and Bobbie Kalman
- 📖 *We're Roaming in the Rainforest: An Amazon Adventure* (Travel the World) by Laurie Krebs and Anne Wilson

CHAPTER 9

- 📖 *From Tadpole to Frog* (Let's-Read-and-Find... Science 1) by Wendy Pfeffer
- 📖 *Frogs and Toads and Tadpoles, Too* (Rookie Read-About Science) by Allan Fowler
- 📖 *National Geographic Readers: Frogs!* by Elizabeth Carney
- 📖 *From Caterpillar to Butterfly* (Let's-Read-and-Find...) by Deborah Heiligman
- 📖 *National Geographic Readers: Great Migrations Butterflies* by Laura F. Marsh
- 📖 *Caterpillars and Butterflies* (Usborne Beginners) by Stephanie Turnbull

CHAPTER 10

- 📖 *Koala* (Life Cycle of A...) by Bobbie Kalman
- 📖 *A Koala Is Not a Bear!* (Crabapples) by Hannelore Sotzek
- 📖 *Rabbits* (Blastoff! Readers: Backyard Wildlife) by Derek Zobel
- 📖 *Rabbits and Raindrops* by Jim Arnosky
- 📖 *The Little Rabbit* by Judy Dunn
- 📖 *The Tale of Peter Rabbit* by Beatrix Potter

CHAPTER 11

- 📖 *Pi-Shu the Little Panda* by John Butler
- 📖 *Endangered Pandas* (Earth's Endangered Animals) by John Crossingham
- 📖 *Tracks of a Panda* by Nick Dowson
- 📖 *Eagles* (Animal Predators) by Sandra Markle
- 📖 *Bald Eagles* (Nature Watch (Lerner)) by Charlotte Wilcox
- 📖 *Challenger: America's Favorite Eagle* by Margot Theis Raven

CHAPTER 12

- 📖 *Deer* (Animals That Live in the Forest) by JoAnn Early Macken
- 📖 *Deer* (Blastoff! Readers: Backyard Wildlife) by Derek Zobel
- 📖 *The Barn Owl* (Animal Lives) by Bert Kitchen
- 📖 *White Owl, Barn Owl* by Michael Foreman and Nicola Davies
- 📖 *There's an Owl in the Shower* by Jean Craighead George
- 📖 *Forest Bright, Forest Night* (Sharing Nature with Children Books) by Jennifer Ward and Jamichael Henterly

CHAPTER 13

- 📖 *Monkeys* (Animals That Live in the Rain Forest) by Julie Guidone
- 📖 *Monkeys and Apes* (Read About Animals) by Dean Morris
- 📖 *If You Give a Mouse a Cookie* (If You Give...) by Laura Joffe Numeroff and Felicia Bond
- 📖 *The Life Cycle of a Mouse* (The Life Cycles Library) by Andrew Hipp

CHAPTER 14

- 📖 *Musk Oxen* (Animals That Live in the Tundra) by Roman Patrick
- 📖 *The Itchy Little Musk Ox* by Tricia Brown and Debra Dubac
- 📖 *Honk, Honk, Goose!: Canada Geese Start a Family* by April Pulley Sayre and Huy Voun Lee
- 📖 *The Goose Man: The Story of Konrad Lorenz* by Elaine Greenstein
- 📖 *Over in the Arctic: Where the Cold Winds Blow* (Sharing Nature with Children Books) by Marianne Berkes and Jill Dubin
- 📖 *The Arctic Habitat* (Introducing Habitats) by Molly Aloian and Bobbie Kalman

CHAPTER 15

- 📖 *Face to Face with Polar Bears* (Face to Face with Animals) by Norbert Rosing and Elizabeth Carney
- 📖 *Polar Bears* by Mark Newman
- 📖 *Where Do Polar Bears Live?* (Let's-Read-and-Find... Science 2) by Sarah L. Thomson
- 📖 *Goats* (Animals That Live on the Farm) by JoAnn Early Macken
- 📖 *Life on a Goat Farm* (Life on a Farm) by Judy Wolfman
- 📖 *Little Apple Goat* by Caroline Church

CHAPTER 16

- 📖 *Face to Face with Penguins* (Face to Face with Animals) by Yva Momatiuk
- 📖 *Emperor Penguin* (Life Cycle of A...) by Bobbie Kalman
- 📖 *National Geographic Readers: Penguins!* by Anne Schreiber
- 📖 *What's It Like to Be a Fish?* (Let's-Read-and-Find... Science 1) by Wendy Pfeffer
- 📖 *Where Fish Go In Winter* by Amy Goldman Koss and Laura J. Bryant
- 📖 *The Life Cycle of Fish* (Life Cycles) by Darlene R. Stille
- 📖 *Eye Wonder: Ocean* by Sue Thornton and Mary Ling
- 📖 *Over in the Ocean: In a Coral Reef* (Sharing Nature With Children) by Marianne Berkes and Jeanette Canyon
- 📖 *A Swim Through the Sea* by Kristin Joy Pratt-Serafini

CHAPTER 17

- 📖 *Face to Face with Whales* (Face to Face with Animals) by Flip Nicklin
- 📖 *Amazing Whales!* (I Can Read Book 2) by Sarah L. Thomson

- *Is a Blue Whale the Biggest Thing There Is?* (Robert E. Wells Science) by Robert E. Wells
- *Squids* (Blastoff! Readers: Oceans Alive) by Colleen A. Sexton
- *Squids* (Under the Sea) by Rake and Jody S.
- *Squirting Squids* (No Backbone! The World of Invertebrates) by Natalie Lunis

CHAPTER 18

- *Discover Science: Animal Disguises* by Belinda Weber
- *Fake Out!: Animals That Play Tricks* (All Aboard Science Reader) by Ginjer L. Clarke and Pete Mueller
- *What Are Food Chains and Webs?* (Science of Living Things) by Bobbie Kalman and Jacqueline Langille
- *Who Eats What? Food Chains and Food Webs* (Let's-Read-and-Find... Science, Stage 2) by Patricia Lauber and Holly Keller
- *DK Readers: The Great Migration* by Deborah Lock
- *Going Home: The Mystery of Animal Migration* (Sharing Nature with Children Books) by Marianne Berkes and Jennifer DiRubbio

QUIZZES

Zoology Quiz #1
Chapters 2 and 3

1. What can you find in the grasslands?

    A. Few trees

    B. Vast grassy fields

    C. Rolling hills

    D. All of the above

2. Lions live in groups called _____.

    A. Herds

    B. Prides

    C. Families

    D. None of the above

3. What does the cheetah's tail help it do?

    A. Find water

    B. Look pretty

    C. Keep is balance

    D. Attract a mate

4. What does an elephant's ears help it do?

    A. Cool down

    B. Communicate with other elephants

    C. None of the above

    D. All of the above

5. The giraffe is the world's _____ animal.

    A. Shortest

    B. Skinniest

    C. Tallest

    D. Fattest

ZOOLOGY QUIZ #2
CHAPTERS 4 AND 5

1. How would you describe desert temperatures?

   A. Cool nights and cool days

   B. Cool nights and hot days

   C. Hot nights and hot days

   D. None of the above

2. The camel can drink up to _____ gallons a minute.

   A. 5

   B. 30

   C. 17

   D. 100

3. The cobra is a _____.

   A. Reptile

   B. Bird

   C. Mammal

   D. Invertebrate

4. What does an omnivore, like the spiny-tailed lizard, eat?

    A. Plants

    B. Insects or other meat

    C. Both

    D. None of the above

5. When does the fennec fox hunt?

    A. During the heat of the day

    B. At night

    C. Whenever it wants

    D. None of the above

Zoology Quiz #3

CHAPTERS 6 AND 7

1. What kind of animals can live on a domestic farm?

    A. Cows

    B. Pigs

    C. Chickens

    D. All of the above

2. What does an herbivore, like the cow, eat?

    A. Plants

    B. Insects or other meat

    C. Both

    D. None of the above

3. Bees are known as the world's _____.

    A. Pollinators

    B. Worst stinging insects

    C. Largest insects

    D. None of the above

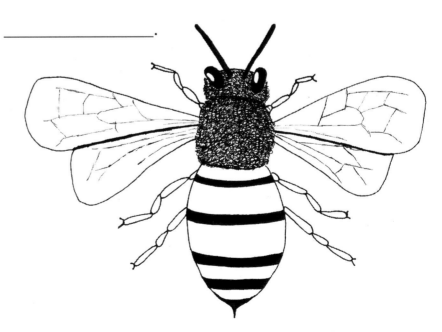

4. Female chickens _____.

    A. Lay eggs

    B. Are known as hens

    C. Are birds

    D. All of the above

5. Spiders are _____.

    A. Insects

    B. Reptiles

    C. Arachnids

    D. Mammals

Zoology Quiz #4
Chapters 8 and 9

1. Where are most of the animals, flowers, and fruits found in the rainforest?

    A. On the forest floor

    B. In the canopy

    C. On top of the trees

    D. In the rivers

2. Where do sloths spend most of their lives?

    A. Hanging upside down

    B. Walking on the forest floor

    C. Swinging through branches

    D. None of the above

3. What do toucans use their beaks for?

    A. To eat

    B. To attract mates

    C. To regulate their temperature

    D. All of the above

4. Poison dart frogs secrete a _____ to cover their bodies.

    A. Toxic slime

    B. Sugar water

    C. Nothing

    D. Orange juice

5. What do blue-morpho butterflies eat?

    A. Small insects

    B. Seeds

    C. Rotten fruit

    D. Leaves

Zoology Quiz #5
Chapters 10 and 12

1. The Australian Brown Mountain Forest has lots of _____ trees.

    A. Oak trees

    B. Pine trees

    C. Eucalyptus trees

    D. Joshua trees

2. Koalas are _____.

    A. Amphibians

    B. Marsupials

    C. Invertebrates

    D. None of the above

3. Why do rabbits dig burrows?

    A. For protection

    B. To have their babies

    C. As shelter from weather

    D. All of the above

4. Owls are _____.

    A. Herbivores

    B. Carnivores

    C. Omnivores

    D. None of the above

5. How often do male deer grow a new set of antlers?

    A. Never

    B. Every month

    C. Every year

    D. Every five years

ZOOLOGY QUIZ #6
CHAPTERS 11 AND 13

1. The _____ lives in the bamboo forest.

    A. Giant panda

    B. Red panda

    C. Golden eagle

    D. All of the above

2. What do pandas eat?

    A. Bamboo

    B. Small animals

    C. Apples

    D. None of the above

3. What makes an eagle an excellent hunter?

    A. Keen eyesight

    B. Powerful talons

    C. Large hooked bill

    D. All of the above

4. Monkeys live in groups called _____.

    A. Prides

    B. Families

    C. Troops

    D. Bands

5. Mice are _____.

    A. Insects

    B. Reptiles

    C. Amphibians

    D. Mammals

Zoology Quiz #7
Chapters 14 and 15

1. Which best describes the arctic?

    A. Very cold winters

    B. Lots of snowfall

    C. Both

    D. None of the above

2. What do musk oxen like to eat?

    A. Lichens

    B. Mosses

    C. Grasses

    D. All of the above

3. Snow geese travel in large groups called _____.

    A. Prides

    B. Flocks

    C. Families

    D. Herds

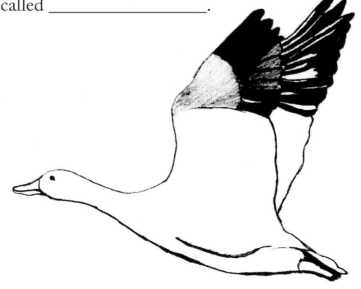

4. Polar bears are _____.

      A. Herbivores

      B. Carnivores

      C. Omnivores

      D. None of the above

5. What is the typical color of the mountain goats of Alaska?

      A. Brown

      B. Black

      C. White

      D. Tan

ZOOLOGY QUIZ #8
CHAPTERS 16 AND 17

1. What part of the earth's surface does the ocean cover?

    A. One-half

    B. Three-quarters

    C. Two-thirds

    D. Three-eighths

2. Penguins _____.

    A. Can't fly

    B. Are excellent swimmers

    C. Are birds

    D. All of the above

3. Where do codfish get the oxygen they need to survive?

    A. From the water

    B. From the air

    C. From the food they eat

    D. None of the above

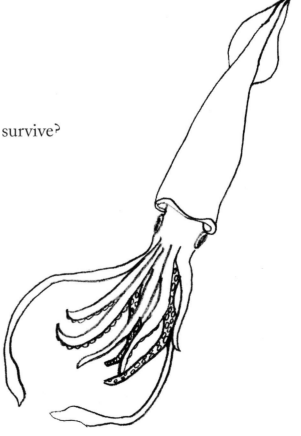

4. Blue whales are _____.

    A. Fish

    B. Mammals

    C. Invertebrates

    D. Birds

5. How many arms do squids have?

    A. 4

    B. 6

    C. 8

    D. 10

Made in United States
North Haven, CT
16 June 2023

37835636R00093